Entrepreneurship, Innovation, and Economic Development

UNU World Institute for Development Economics Research (UNU-WIDER) was established by the United Nations University as its first research and training centre and started work in Helsinki, Finland, in 1985. The purpose of the institute is to undertake applied research and policy analysis on structural changes affecting developing and transitional economies, to provide a forum for the advocacy of policies leading to robust, equitable, and environmentally sustainable growth, and to promote capacity strengthening and training in the field of economic and social policy-making. Its work is carried out by staff researchers and visiting scholars in Helsinki and via networks of collaborating scholars and institutions around the world.

United Nations University World Institute for Development Economics Research
(UNU-WIDER)
Katajanokanlaituri 6 B, 00160 Helsinki, Finland
www.wider.unu.edu

UNU-MERIT is a joint research and training centre of United Nations University (UNU) and Maastricht University, the Netherlands. The joint institute was created on 1 January 2006 following the integration of the former UNU Institute for New Technologies (INTECH) in Maastricht, and the Maastricht Economic Research Institute on Innovation and Technology, MERIT, at Maastricht University. UNU-MERIT provides insights into the social, political, and economic factors that drive technological change and innovation. The Centre's research and training programmes address a broad range of policy questions relating to the national and international governance of science, technology, and innovation, with a particular focus on the creation of, diffusion of, and access to knowledge.

United Nations University Maastricht Economic and Social Research Institute
on Innovation and Technology
(UNU-MERIT)
Keizer Karelplein 19, 6211 TC Maastricht, the Netherlands
www.merit.unu.edu

Entrepreneurship, Innovation, and Economic Development

Edited by

Adam Szirmai, Wim Naudé, and Micheline Goedhuys

A study prepared for the World Institute for Development Economics Research of the United Nations University (UNU-WIDER) and Maastricht Economic and Social Research Institute on Innovation and Technology (UNU-MERIT)

OXFORD
UNIVERSITY PRESS

OXFORD
UNIVERSITY PRESS

Great Clarendon Street, Oxford OX2 6DP

Oxford University Press is a department of the University of Oxford.
It furthers the University's objective of excellence in research, scholarship,
and education by publishing worldwide in

Oxford New York

Auckland Cape Town Dar es Salaam Hong Kong Karachi
Kuala Lumpur Madrid Melbourne Mexico City Nairobi
New Delhi Shanghai Taipei Toronto

With offices in

Argentina Austria Brazil Chile Czech Republic France Greece
Guatemala Hungary Italy Japan Poland Portugal Singapore
South Korea Switzerland Thailand Turkey Ukraine Vietnam

Published in the United States
by Oxford University Press Inc., New York

First published 2011

British Library Cataloguing in Publication Data

Data available

Library of Congress Cataloging in Publication Data

Data available

Typeset by SPI Publisher Services, Pondicherry, India
Printed in Great Britain
on acid-free paper by
MPG Books Group, Bodmin and King's Lynn

ISBN 978–0–19–959651–5

1 3 5 7 9 10 8 6 4 2

Foreword

Entrepreneurship and innovation are two of the most pervasive concepts of our times. A century ago Joseph Schumpeter made a fundamental contribution to our understanding of economic development by highlighting the necessity of innovation and the key role of the entrepreneur as innovator. There are, however, still important gaps in our understanding of the interactions between entrepreneurship and innovation, particularly in developing countries where the challenges of erasing disparities, catching up with advanced economies, achieving the Millennium Development Goals, and mitigating and adapting to man-made and natural disasters, including climate change, remain paramount.

This book is an attempt to fill these gaps. It focuses on the entrepreneurship–innovation–development nexus, drawing heavily on empirical evidence from developing countries. The cross-country and individual country experiences analysed here cover countries as diverse as Ethiopia, India, Turkey, and Vietnam. The book also examines lessons to be learned from advanced economies such as Finland. The chapters of the book were first presented at a workshop held in Maastricht on 30–31 October 2008. This workshop was jointly organized by UNU-MERIT and UNU-WIDER, a collaborative effort that aimed to bring the best of the complementary research expertise and networks of these two organizations to bear on complex social phenomena such as entrepreneurship and innovation.

Three important aspects of the entrepreneurship–innovation–development nexus are elaborated in this book. These aspects are the impact of entrepreneurship and innovation on growth and development, the determinants of innovative performance in developing countries, and the role played by the institutional environment. A key message is that entrepreneurial innovation, whether through small firms, large national firms, or multinational firms, is often vibrant in developing countries. But unfortunately it is not realizing its full potential. Many developing countries do not have conducive environments for doing business or pursuing innovation. They need to reform their institutions to create more positive incentives. Nevertheless, some entrepreneurs do succeed in overcoming these obstacles. They can even actively contribute towards institutional development, which in itself can be seen as an act of innovation.

It is clear from this book that there is considerable scope for increasing entrepreneurial innovation in developing countries. Hence, important and exciting challenges continue to face researchers and policy makers.

Luc Soete, Director of UNU-MERIT, and Finn Tarp, Director of UNU-WIDER

Acknowledgements

This book is the result of a collaborative research effort by two 'sister' institutions of the United Nations University, UNU-MERIT (Maastricht Economic and Social Research Institute on Innovation and Technology) and UNU-WIDER (World Institute for Development Economics Research). For us, and hopefully for all of the participants in this research effort, this collaboration has been an enriching and highly rewarding experience. This has been partly due to the fact that MERIT and WIDER's research agendas and resident expertise converged on the topic of innovation and entrepreneurship in the development process. Hence, we could support one another's research and also learn a lot from each other.

More importantly, however, it has been an exceptional experience because we were amply supported by so many enthusiastic colleagues and friends. In addition to the scholars who contributed chapters and willingly revised and even rewrote parts of their chapters in light of the broader aims of the book, we are extremely grateful to Luc Soete (director of MERIT) and Tony Shorrocks and Finn Tarp (former and current directors of WIDER) for their broad support. We thank Eveline in de Braek and Lisa Winkler for their efficient and professional project support, particularly during the hosting of the project workshop in Maastricht in October 2008. Luc Christiaensen, Lorraine Telfer-Tairainen, and Lisa Winkler gave much of their time and expertise in terms of assessment and editorial support in the preparation of the manuscript. Three anonymous referees provided us with valuable comments, criticisms, and feedback.

Last but not least, we wish to gratefully acknowledge the financial assistance that UNU-WIDER received from the Finnish Ministry of Foreign Affairs to the research project on 'Promoting Entrepreneurial Capacity'. UNU-WIDER also wishes to gratefully acknowledge contributions to its research programmes by the governments of Denmark (Royal Ministry of Foreign Affairs), Finland (Ministry for Foreign Affairs), Sweden (Swedish International Development Cooperation Agency—Sida), and the United Kingdom (Department for International Development—DFID).

Adam Szirmai, Wim Naudé, and Micheline Goedhuys

Maastricht, the Netherlands and Helsinki, Finland
July 2010

Contents

List of Tables

List of Abbreviations

ABLE	Association of Biotechnology Led Enterprises
ANPROTEC	National Advanced Technology Enterprise Promoter Entity
BPO	Business Process Outsourcing
BRIC	Brazil, Russia, India, and China
BT	Biotechnology
CEE	Central and Eastern Europe
CEO	Chief Executive Officer
CES	Constant Elasticity of Substitution
CII	Confederation of Indian Industry
CIS	Community Innovation Survey
CSA	Central Statistical Agency
DST	Department of Science and Technology
EC	European Commission
ERP	Enterprise Resource Planning Product
EU	European Union
FDI	Foreign Direct Investment
FICCI	Federation of Indian Chamber of Commerce and Industry
FOE	Foreign-owned Enterprise
GCI	Growth Competitive Index
GDP	Gross Domestic Product
GEM	Global Entrepreneurship Monitor
GPT	General Purpose Technology
GRET	Groupe de Recherche et d'Échanges Technologiques
HNI	High Networth Individual
ICT	Information and Communication Technology
IIM	Indian Institute of Management
IIT	Indian Institutes of Technology
ILO	International Labour Organization

IP	Internet Protocol
IP	Intellectual Property
IPR	Intellectual Property Right
IT	Information Technology
ITRI	Industrial Technology Research Institute
IV	Instrumental Variable
KOSGEB	Turkish Small and Medium-sized Industry Development Organization
LAN	Local Area Network
LSE	London Stock Exchange
MNC	Multinational Corporations
MNE	Multinational Enterprise
MOST	Ministry of Science and Technology
MSE	Micro and Small-sized Enterprises
NASDAQ	National Association for Securities Dealers Automated Quotations
NASSCOM	National Association of Software Service Companies
NBIA	National Business Incubation Association
NEN	National Entrepreneurship Network
NGO	Non-governmental Organization
NIS	National Innovation System
NIT	Núcleos de Inovação Tecnológica
NSTEDB	National Science and Technology Entrepreneurship Development Board
NYSE	New York Stock Exchange
OECD	Organisation for Economic Co-operation and Development
OIZ	Organized Industrial Zone
OLS	Ordinary Least Squares
PACTI	Industry Technological Training Support Programme
PE	Private Equity
PNI	National Incubation Support Programme
POE	Privately Owned Enterprise
PPP	Purchasing Power Parity
R&D	Research and Development
RBI	Reserve Bank of India
SAP	Systems, Applications, and Products in Data Processing
S&T	Science and Technology
SEBRAE	Micro and Small Business Support Services

SME	Small and Medium-sized Enterprise
SOE	State-owned Enterprise
SSA	Sub-Saharan Africa
STEP	Science and Technology Entrepreneurs Park
TBF	The Bamboo Factory
TBI	Technology Business Incubators
TCS	Tata Consultancy Services
Tekes	National Technology Agency of Finland
TEKMER	Technology Development Centres
TePP	Technopreneur Promotion Programme
TiE	The Indus Entrepreneurs
TIME IS	Technology Innovation Management and Entrepreneurship Information Service
TSI	Turkish Statistical Institute
TÜSİAD	Turkish Industrialists' and Businessmen's Association
UN	United Nations
US	United States
VC	Venture Capital
WB	World Bank
YB	Young Business

Notes on the Contributors

İ. Semih Akçomak holds a PhD in economics of Maastricht University. His main research interests include social and human capital, institutions, innovation policy, and regional economics. He currently works as a researcher at the Department of Economics, Maastricht University and UNU-MERIT, and he is also affiliated with the Netherlands Bureau for Economic and Policy Analysis (CPB).

Alice H. Amsden is Barton L. Weller Professor of Political Economy at the Massachusetts Institute of Technology, Department of Urban Studies and Planning, International Development Group. Her research focus is on economic and industrial development. Selected recent publications include *Beyond Late Development: Taiwan's Upgrading Policies* co-authored with Wan Wen Chu (MIT Press 2003) and *The Rise of the Rest: Challenges to the West from Late-Industrializing Countries* (Oxford University Press 2003).

Suma Athreye is Professor of International Strategy at Brunel University in the UK. Her research focuses on the impact of technology markets on business and public policy issues. A second strand of her research is focused on the use of technology by Indian firms and the impact of technological and regulatory change on Indian businesses. She has won several grants for research which have also been recognized by the national and international media.

David B. Audretsch is a distinguished professor and the Ameritech Chair of Economic Development and Director of the Institute for Development Strategies at Indiana University. He is also the Director of the Max Planck Institute of Economics in Jena, Germany. Professor Audretsch is ranked as the twenty-first most cited scholar in economics and business, 1996–2006. He was awarded the 2001 International Award for Entrepreneurship and Small Business Research by the Swedish Foundation for Small Business Research. In 2008, he received an honorary doctorate degree from the University of Augsburg. His book *The Entrepreneurial Society* was published by Oxford University Press in 2007.

Elif Bascavusoglu-Moreau is a research associate at the Innovation and Entrepreneurship Group at Imperial College Business School, where she is working on the ESRC-funded research project 'Open Innovation and Servicisation Strategies' with Professor Bruce Tether. She holds a PhD in international economics of the University of Paris I Panthéon-Sorbonne, and has worked as a post-doctoral research fellow at the Innovation, Knowledge and Development (IKD) Centre at the Open

University on improving the understanding of the drivers and dynamics of innovation. Her research interests are technology diffusion, knowledge spillovers, and IPRs.

Mulu Gebreeyesus is currently Senior Research Fellow at UNU-MERIT. He was previously Research Fellow and Head of the Poverty and Sectoral Development Directorate at the Ethiopian Development Research Institute (EDRI). His research interests include firm competitiveness and technological capabilities, trade openness, and microenterprises performance. He holds a PhD in economics from the University of Gothenburg, Sweden.

Micheline Goedhuys is a research fellow at UNU-MERIT and at Maastricht University. She holds a PhD in economics from the University of Leuven, Belgium. She has worked as a consultant for several international organizations such as the World Bank and the ILO. She has published widely in international journals on topics related to entrepreneurship, small business development, and innovation and firm performance in developing countries, with a particular focus on Africa. She has recently been guest editor of *Structural Change and Economic Dynamics* and *the European Journal of Development Research* for a special issue on 'Micro-evidence on Innovation and Development'.

Gerard de Groot is Director of the Development Research Institute (IVO) of Tilburg University. He has experience in research and capacity building projects, mainly in South Asia and Sub-Saharan Africa. His main research interest is commodity chain analysis with emphasis on the aspects of value distribution and impact analysis, especially with respect to non-profit organizations.

Job de Haan is Associate Professor of Operations Management at the Department of Organization and Strategy at the University of Tilburg. He has been involved in various research projects in the areas of international production and logistics systems and supply chains in which developing countries participate. His current research focuses on lean logistics service providers and on small scale innovation.

Sunil Mani is Professor, Planning Commission Chair in Development Economics at the Centre for Development Studies in Trivandrum, Kerala, India. His research interests include new indicators for measuring innovation, policy instruments to stimulate innovation at the enterprise level, and the telecommunications industry. His recent publications include an edited book with Franco Malerba on *Sectoral Systems of Innovation and Production in Developing Countries: Actors, Structure and Evolution* (Edward Elgar 2009) and *Innovation Capability in Developing Countries: A Study of the Telecommunications Industry* (Edward Elgar 2010).

Wim Naudé is Senior Research Fellow at UNU-WIDER. A graduate of the University of Warwick, he has been Research Officer at the University of Oxford, and Director of Research at North-West University, South Africa. He has also served as a board member of the International Council for Small Business. His research focuses on global economic development, regional development, and entrepreneurship. He is editor of the book *Entrepreneurship and Economic Development,* published by Palgrave Macmillan. He has been guest editor of a special issue of *Small Business Economics,* devoted to entrepreneurship and economic development, and of the *European Journal of Development Research*, dealing with female entrepreneurship in developing countries.

Mark Sanders is Assistant Professor in Economics at Utrecht University, the Netherlands, and Research Fellow of the Max Planck Institute of Economics in Jena, Germany. His research interests include innovation economics, entrepreneurship, and economic development with a special focus on the role of entrepreneurial activity in the introduction of renewable energy technologies. Mark Sanders is teaching a wide variety of topics in economics at undergraduate and graduate levels. His publications include a PhD thesis on the impact of innovation on skilled labour demand and papers on the role of entrepreneurs in the innovation process.

Erik Stam is Associate Professor at the Utrecht School of Economics, Utrecht University. He is AIM Research Fellow at the University of Cambridge and is also affiliated with the Scientific Council for Government Policy (The Hague), and the Max Planck Institute of Economics in Jena, Germany. He received his MSc in economics and geography as well as his PhD from Utrecht University, the Netherlands.

Adam Szirmai is Professorial Fellow at UNU-MERIT and Professor of Development Economics at the Maastricht Graduate School of Governance of Maastricht University. His research focuses on technology productivity and growth in manufacturing in developing countries. His textbook on development studies *The Dynamics of Socio-economic Development* was published by Cambridge University Press in 2005. A second edition is in preparation.

Otto Toivanen is the Director of the Helsinki Center of Economic Research. Earlier he was a professor at the Helsinki School of Economics, and has been a visiting scholar at MIT and UC Berkeley. His research interests include industrial organization and economics of innovation. He has consulted for the Finnish government and the EU Commission on innovation policy.

André van Stel is a senior researcher at EIM Business and Policy Research in Zoetermeer and at the Amsterdam Center for Entrepreneurship of the University of Amsterdam. He is also a research fellow at the Tinbergen Institute, Amsterdam. He holds a PhD in economics and a Master's Degree in econometrics, both from Erasmus University Rotterdam.

Jaap Voeten, a development economist by education, has been a PhD development economist by education, *Jaap Voeten* is a PhD researcher at the Development Research Institute (IVO) at Tilburg University since 2007, specializing in innovation, small producers, and poverty alleviation issues framed in contemporary theories on innovations systems, evolutionary economic theory, and new institutional economics. He has a long-term project, policy development, and research experience in Vietnam on entrepreneurship and small business development.

Part I
Introduction

1

Entrepreneurship, Innovation, and Economic Development: An Overview

Adam Szirmai, Wim Naudé, and Micheline Goedhuys

1.1 Purpose of the book

Economic development requires sustainable and shared increases in per capita income accompanied by changes in the structural composition of an economy towards higher value added goods and more efficient production methods. Entrepreneurs can contribute to economic development by facilitating the reallocation of resources from less to more productive uses (Acs and Storey 2004), by performing 'cost-discovery', 'gap-filling', and 'input-completing' functions in the economy (Leibenstein 1968; Hausmann and Rodrik 2003) and by supporting structural change (Lewis 1954; Gries and Naudé 2010). These roles have recently been the subject of a growing literature (see e.g. Minniti and Naudé 2010; Naudé 2010a, 2010b, 2010c).

A neglected function in this literature is the potential role of entrepreneurs as innovators in developing countries. Joseph Schumpeter pointed out a century ago that entrepreneurs are often innovators, bringing new goods and technologies to markets, opening up new markets, processes, and ideas, and commercializing new knowledge. But, it is often mistakenly suggested that innovation by entrepreneurs is less important for growth in low-income developing countries than in more advanced economies.[1]

A substantial literature has been devoted to understanding the conditions under which entrepreneurs innovate, and the nature and evolution of national systems of innovation (see e.g. Lundvall 1992; Nelson 1993). By and large, however, this literature has been concerned with the process of innovation and its dynamics in advanced economies. Research at the intersection of the fields of entrepreneurship, innovation, and development is still in its infancy.

In this book we provide perspectives on three broad questions pertaining to innovation and entrepreneurship in developing countries. How does innovation impact on development? How and under what conditions do entrepreneurs in developing countries innovate? And, what can be done to support innovation by entrepreneurs in developing countries? The book is divided into three parts: this introductory part, followed by Part II dealing with the impact and determinants of innovation by entrepreneurs in developing countries, and Part III, focusing on policies and institutions for innovation. But first, we need to define what we mean by entrepreneurship and innovation and provide a brief background on what is known about the impact and determinants of innovation, and the importance of the policy and institutional environment for fostering innovation.

1.2 Concepts and definitions

1.2.1 *Entrepreneurship*

The discipline of entrepreneurship generally studies the why, when, and how of opportunity creation, recognition, and utilization. Hence, a widely quoted definition of entrepreneurship is that it is about the 'discovery and exploitation of opportunities' (Shane and Ventakaraman 2000).[2] Baumol (1990) recognized that not all opportunity exploitation will necessarily be in society's best interest, and he defined entrepreneurs very broadly as 'persons who are ingenious and creative in finding ways that add to their own wealth, power, and prestige' (ibid.: 987). Thus entrepreneurial talent can be allocated in ways that retard economic development. Although we recognize that the 'reward structure' of a society can lead to such a destructive allocation of entrepreneurial talent, we will in this book be focusing on productive entrepreneurial activity. This consists of the creation, recognition, and utilization of *positive* opportunities within existing firms (or through creation of new firms) in such a way that involves 'innovation'—or the provision of 'new combinations'.

Before defining what is meant by this 'innovation' (see section 1.2.2) we should note that three main conceptual approaches to entrepreneurship can be distinguished in the literature. The first approach focuses on the entrepreneurial function, the second on the performance of enterprises and the third on owner-operated enterprises.

The functional perspective is concerned with the dynamic actors that make key decisions on investment, production, innovation, location, or research and development. This conception of entrepreneurship is broader than that of entrepreneurs who run their own businesses. It also includes managers of multinational firms, state enterprises, or non-profit organizations and a variety of dynamic intrapreneurs within organizations. In this perspective, entrepreneurship is a psychological trait referring to dynamism, creativity, and originality. As in the early Schumpeterian tradition, the difference between entrepreneurial behaviour and innovation is blurred: entrepreneurial behaviour is innovative behaviour. If one is not innovative, one is not entrepreneurial.

The second strand of research focuses on the firm as the key economic actor. The firms studied include owner-operated firms, incorporated joint stock companies, state-owned firms, joint ventures, and subsidiaries of multinationals. The firms are the units that make the key decisions on investment, on branching into new activities or sectors or relocating to other countries. There is by now a large literature on firm-level behaviour in developing countries, examining firm characteristics, including their economic performance, their innovative performance, their capabilities and their business strategies (e.g. Goedhuys et al. 2008; Goedhuys and Sleuwaegen 2010).

The third research tradition deals with an important sub-set of firms, namely owner-operated enterprises. The entrepreneur is the person who is both owner and actively involved in running his/her own business. Here the focus is often on small and medium-sized enterprises (SMEs) and self-employment, as exemplified by many papers in this collection. Like the second approach, this research tradition tries to distinguish between high potential, innovative firms that survive and grow and stagnant firms that barely survive or exit the market. One sub-category of firms which receives special attention is that of start-ups, especially in the work associated with the Global Entrepreneurship Monitor (GEM), which provides estimates of start-up rates across countries. Though the emphasis of this research tradition is on SMEs, we should not forget that very large companies are sometimes also run by their entrepreneurial owners (see Amsden's Chapter 3 in this book).

1.2.2 *Innovation*

Following the Oxford Handbook of Innovation (Fagerberg, Mowery, and Nelson 2005), the concept of innovation refers to the putting into practice of inventions.[3] A narrow strictly technological approach focuses specifically on product and process innovations, or technological innovation, often said to be the result of knowledge-intensive (technological) entrepreneurship. A broader approach refers to innovation as the development of new products, new processes, new sources of supply, but also to the exploitation of new markets and the development of new ways to organize business. One can distinguish between more incremental innovations and more radical innovations. It is important to note that innovation does not only refer to the first introduction of novelty by a first mover, but also to the spread of the innovation to other actors in the economy.

Versions of these concepts of innovation are used by the authors in this book. For instance Toivanen (Chapter 8) and Mani (Chapter 9) focus on knowledge-intensive entrepreneurship with an emphasis on product and process innovations. Stam and van Stel (Chapter 4), Voeten, de Haan, and de Groot (Chapter 5), Gebreeyesus (Chapter 6), and Bascavusoglu-Moreau (Chapter 7) see innovation more broadly as including the exploitation of new markets, new forms of organization, and new sources of supply. Audretsch and Sanders discuss the impacts of general purpose technologies. Here, clusters of innovations result in the emergence of technologies such as electricity or information and communication technology (ICT) which have a pervasive influence across all sectors of the economy (see also Lipsey, Carlaw, and Bekar 2005).

An important distinction in the innovation literature is between innovations that are new to the world, innovations that are new to the domestic market or innovations that are new to the firm (Fagerberg 2005). Innovations that are new to the world are primarily found in the advanced economies. They are based on research and development at the frontiers of global knowledge. In developing countries far removed from the international technological frontier, innovations will tend to be new to the market or new to the firm. Innovations new to the market in developing countries refer to the international diffusion and absorption of technology. The domestic firm introduces innovations which have already been developed elsewhere, but which are new to the market in their own country. Innovations new to the firm refer to knowledge flows within the domestic economy. The innovation is already present in the market, but is now adopted by a given firm. This last concept of innovation comes closest to the Rogerian concept of innovation (Rogers 2003).

What is new to the firm may not be very innovative in any objective sense. It may be simply introducing a machine for moulding handles of kitchen knives, rather than doing it by hand or introducing a new oven for hardening ceramics (as discussed by Voeten, de Haan, and de Groot). This means that some kinds of innovation that are new to small firms in developing countries may coexist with stagnant economies and increasing technology gaps relative to the international frontier. Our primary interest is in the kinds of innovative behaviour that promote economic dynamism and catch up at country-level.

Like entrepreneurship, innovative performance has been measured in a variety of ways, using patents, trademarks, R&D inputs, and other secondary indicators such as publications or citations. Since the 1980s, increasing use has been made of innovation surveys amongst firms. In innovation surveys, firms are asked whether they have introduced innovations. The main focus in most innovation surveys is on technological innovations resulting in new products or new production processes. In later generations of the surveys, a separate category of organizational and marketing innovations is distinguished. Starting with the European Community Innovation Surveys innovation surveys have spread to the developing world, in particular to Latin America, but also to Asia and Africa.

In this book, different approaches to measurement are represented. The chapters by Gebreeyesus and Bascavusoglu-Moreau are based on survey data and measure innovation using survey questions about product, process, and marketing innovations at firm-level. The chapter by Voeten, de Haan, and de Groot uses a qualitative case study approach to analyse the introduction of new products, new processes, and the opening of new markets in clusters of small-scale producers in Vietnamese villages. The chapters by Mani and Toivanen use a variety of secondary indicators of innovative performance such as R&D, scientific publications, or patents to measure innovation trends at the macro-level.[4]

1.2.3 *Innovation and entrepreneurship*

If one follows the Schumpeterian tradition in describing the search for new combinations as the key characteristic of entrepreneurship, the distinction between entrepreneurship and innovation is hard to make. Is there really any difference between entrepreneurship and innovation? In a book that attempts to link two research traditions on respectively entrepreneurship and innovation, we need to clarify the relationships between these concepts.

In the Schumpeterian tradition, the entrepreneur is the hero of dynamic capitalism. The entrepreneur typically creates new combinations: new products, new markets, new materials, and new forms of organization (Schumpeter 1934). Entrepreneurship and innovation are almost synonymous. However, on the basis of Schumpeter's later work (Schumpeter 1943), one needs to distinguish between two types of competition, often referred to as Schumpeter Mark I and Schumpeter Mark II. In Mark I, it is the entrepreneurs and the smaller enterprises that are responsible for innovation. In Schumpeterian competition Mark II large oligopolistic firms predominate. Innovation takes place in R&D laboratories and the entrepreneurial function has become routinized in a managerial bureaucracy. Innovation has become divorced from entrepreneurship.

The disadvantage of the identity of entrepreneurship and innovation in Mark I is that it does not allow us to distinguish different types of entrepreneurship: innovative entrepreneurship, non-innovative entrepreneurship, stagnant entrepreneurship, destructive entrepreneurship (Baumol 1990), high-growth entrepreneurship, or survival entrepreneurship (see also Naudé

2010a). Also, it does not recognize that much innovation is generated in firms that are run by managers rather than entrepreneurs. From a developmental perspective, it therefore makes analytic sense to distinguish entrepreneurship and innovation as distinct key forces in development. Some entrepreneurs are much more innovative than others and firms managed and owned by entrepreneurs are not the only sources of innovation.

1.3 Background

As we indicated in section 1.1, the purpose of this book is to provide new perspectives on three questions pertaining to entrepreneurship and innovation in developing countries: How does innovation impact on development? How, and under what conditions, do entrepreneurs in developing countries innovate? And, what can be done to support innovation by entrepreneurs in developing countries? Before we summarize the answers of the various chapters to these questions, we will first provide some background against which to evaluate the contribution of the various chapters that follow.

1.3.1 *The impact of innovation on development*

Innovation is central to modern theories of growth and development (Verspagen 2005). Along with the traditional factors such as costs, technological product, and process, innovations have become the key to competitiveness and business success. Competition in the global economy has increasingly become knowledge-based. Even in supposedly traditional economic sectors such as textiles, leather, or food processing innovation and technological advance has become the key to growth (c.f. Mytelka 1999). The same holds for service sectors such as distribution and retailing, financial services, and ICT services. Innovation is also intimately tied up with changes in the structure of the economy, technological upgrading in production, and moving to higher value added activities in global value chains.

Technological change is embodied in new generations of machinery and equipment and new generations of better educated workers. There are also disembodied advances in product and process technology, which result from formal and informal investment in R&D, capabilities, and on-the-job learning. Embodied and disembodied technological change raises total factor productivity—which has been found to explain more than half of the variation in economic growth rates between countries (Helpman 2004). But it not only raises the quantity of economic output, but also the quality and nature of what is produced. It results in an ever wider range of new goods and services.

This powerful impact of technological change is aptly described by Lipsey, Carlaw, and Bekar (2005: 5):

> People living in the first decade of the twentieth century did not know modern dental and medical equipment, penicillin, bypass operations, safe births, control of genetically transmitted diseases, personal computers, compact discs, television sets, automobiles, opportunities for fast and cheap worldwide travel, affordable universities, central heating, air conditioning... technological change has transformed the quality of our lives.

Both endogenous growth theory and evolutionary growth theory emphasize that the traditional factors of production such as labour or capital are subject to diminishing returns, while investment in knowledge has increasing returns due to positive externalities and knowledge spillovers between economic actors (e.g. Romer 1990). Endogenous growth theory argues that the most advanced economies with their superior systems of innovation profit more from investment in knowledge than less advanced economies. First, R&D efforts and scientific research are still overwhelmingly concentrated in the most advanced economies (Szirmai 2008, 2011). Next, the flow of knowledge and technology from first movers to followers is very rapid, so that innovations quickly diffuse throughout the economy. Endogenous growth theory thus helps us understand the process of divergence in per capita incomes between rich and poor countries in the world economy.

However, innovation and technological advance can also result in accelerated catch-up in developing countries. What endogenous growth theory fails to capture is the fact that in an increasingly unequal world economy, several developing countries have experienced rapid economic catch-up. They were able to absorb and creatively adapt international technological knowledge to achieve accelerated growth. Gerschenkronian and evolutionary growth theories argue that latecomer economies may profit from the advantages of technological backwardness. They can benefit from global diffusion of technology. They can access new technologies without bearing all the costs and risks of investment in new knowledge. Amsden's chapter (Chapter 3) argues that privately owned domestic firms in East Asia were better at adopting and absorbing technologies from advanced economies than foreign-owned firms. Stam and van Stel (Chapter 4) highlight how the adoption of foreign technology provides entrepreneurs with a potential to create new markets and contribute to structural change and self-discovery.

Whether developing countries are able to profit from the advantages of technological backwardness clearly depends on their social capabilities and absorptive capacities. Hence, importantly for developing countries, innovation does not only refer to the development of new products or processes, but also to the capacity to creatively absorb technology. If the absorptive capacities of a country are sufficiently developed, very rapid economic growth in a technologically backward country is not a miracle, it is the norm (Szirmai 2005, 2011).

It is here where the entrepreneur makes his or her entrance. Traditional macroeconomic growth theory is a black box which relates inputs and outputs. The study of entrepreneurship opens this black box and allows us to analyse the characteristics and choices of different types of firms and entrepreneurs that are responsible for capital accumulation, hiring of workers, structural change, and the development or adoption of new technologies (for a modelling approach see Gries and Naudé 2010 and Chapter 2 by Audretsch and Sanders). The entrepreneurs are the actors that respond to opportunities, threats, uncertainties, constraints, and incentives emanating from the economic environment in which they operate. This puts entrepreneurship at the heart of economic growth, development, and catch-up.

By innovating and commercializing inventions and by adopting innovations developed by others, developing country entrepreneurs affect the rate of technological change and the structural transformation of the economy. Entrepreneurs, commercializing technology, often through creation or expansion of firms, apply and spread technology in a way which raises total factor productivity. The creativity, capabilities, dynamism, and innovativeness of the entrepreneurs in a country are important aspects of the absorptive capacity, which is such a distinctive characteristic of successful development experiences. How entrepreneurs perform this function will vary across various stages of a country's development.

It is important to take the context of developing countries into account. In catch-up countries, innovative entrepreneurs initially focus on delivering incremental improvements to existing foreign designs, rather than the risky development of products and technologies that are new to the world (see Athreye's Chapter 11 in this book, and Hobday and Perini 2009). In later stages of development, they will gradually shift to innovations which are more novel in a global perspective. Thus, the challenges faced by the entrepreneurs will also change in the course of economic development.

1.3.2 *Determinants of innovation in developing countries*

The chapters in this book clearly illustrate that some entrepreneurs are more innovative than others and entrepreneurs in some countries are more innovative than in other ones. Why is this the case and what are the determinants of innovative performance? Some of the explanations are found in characteristics of the firm or of the entrepreneur such as education and experience of the entrepreneur, firm size and age, or organizational culture. Such explanations are examined in the chapters by Gebreeyesus, Bascavusoglu-Moreau, and Voeten, de Haan, and de Groot. But we are also interested in the extent to which market conditions, policies, and the institutional environment can promote or hinder innovative behaviour.

The fundamental reason *why* entrepreneurs innovate has been answered already long before Schumpeter. They are driven by profit motives. Adam Smith's important insight was to realize that although entrepreneurs act in pursuit of their own profits, they may generate benefits to the broader society in the process. His further insight was that there is a link between the degree to which the entrepreneur will engage in technical innovation, specialization, and the size and functioning of the market. Markets can thus be seen as important drivers of growth and development.

In the poorest developing countries, markets unfortunately fail to fulfil this role. They are hamstrung in various ways, many of them already analysed by Adam Smith in his *Wealth of Nations*. Developing country markets are often small, fragmented, and imperfect due to lack of infrastructure, low per capita incomes, misguided policies, and institutional constraints. The political stability, predictability and transparency, peace and other institutional prerequisites for the functioning of markets are often absent. With fragmented, small and uncertain markets there is insufficient incentive for entrepreneurs to innovate. Where markets are restricted because of barriers to trade (either natural barriers such as lack of infrastructure or man-made barriers), it is difficult for innovations to spread. Throughout the ages international trade has exposed traders and merchants to new ideas and technologies. This is one of the reasons why trade functions as an engine of growth. Where markets are restricted by inappropriate regulations or strangled by predatory governments or monopolies, there is no incentive for entrepreneurs to introduce innovations that are new to the firm. Where inappropriate property rights and weak contract enforcement makes returns to innovative activity risky, there will be little incentive for entrepreneurs to invest in innovations new to the domestic market or new to the world.

1.3.3 *Policies and institutions for innovation in developing countries*

While the broadening of the market is one of the necessary conditions for innovation as outlined in the previous section, it will not be sufficient. The reason is that innovation is increasingly knowledge- and skill-intensive. Because of the positive externalities inherent in investment in knowledge, technological advance, and human capital, public policy has been increasingly recognized as having an important complementary role to play in fostering entrepreneurial innovation. Innovation requires not only highly knowledgeable, experienced, and skilled entrepreneurs, but also highly skilled labourers. Thus, educational policies and capability building come into the picture. In the absence of government interventions and policies, the operation of markets results in underinvestment in knowledge and innovation. Nowadays, 'innovation policy' and 'national innovation systems' have become a standard part of the economic growth discourse in both advanced economies and developing economies (Lundvall 1992; Nelson 1993).

In fact the very concept of 'innovation policy' had its origins in the United States,[5] widely seen as one of the most entrepreneurial economies in the world. As Lazonick has pointed out (Lazonick 2008: 2–3):

> In *all* the advanced economies over the past century, *first and foremost the United States* where the ideology of 'free market' entrepreneurialism is most virulent, successful entrepreneurship has depended heavily upon government investment in the knowledge base, state sponsored protection of markets and intellectual property rights, as well as state subsidies to support business investment strategies.

In developing countries, the benefits of innovation by entrepreneurs depend on the characteristics of the system of innovation within which they are embedded. The better the system of innovation, the more a developing country will be able to tap into global technology, the more rapidly knowledge will circulate within the domestic economy, and the more rapidly the economy will embark on the process of technological upgrading. The weaker the system of innovation, the less the efforts of individual entrepreneurs will contribute to accelerated economic development and catch-up (Goedhuys and Srholec 2010). Stam and van Stel (Chapter 4) indeed observe a greater contribution of dynamic entrepreneurship to economic growth in more advanced economies than in developing countries, where low levels of human and financial capital, the absence of a sound firm size distribution, and weak institutional frameworks limit the contribution of entrepreneurship to growth. The interplay between market development, systems of innovation and public science, technology, and innovation policies is an important theme of this book—particularly taken up in Part III.

1.4 Contributions of the book

In this section we provide an analytic summary of the contributions in the book focusing on the key questions pertaining to entrepreneurship and innovation in developing countries.

1.4.1 *Impacts and determinants of innovation in developing countries*

In an increasingly globalized world economy, some developing countries have experienced very rapid economic catch-up. They are able to absorb and creatively adapt international technological knowledge to achieve accelerated growth. In subsequent stages of development they become innovators in their own right. Their absorptive capacities distinguish these countries from developing countries that are becoming marginalized in the global economy. The absorptive capacities depend among others on the nature of entrepreneurship in these countries. Several of the chapters in Part II of this book analyse the roles and impacts of different types of entrepreneurship in the process of economic development. The different chapters discuss a variety of actors: owner-operators of small and medium-sized firms, larger private enterprises, state-led or state dominated enterprises, or subsidiaries of foreign multinational firms. One of the interesting questions concerns the relative importance of the different kinds of actors as drivers of growth.

In their chapter 'Technological Innovation, Entrepreneurship, and Development' (Chapter 2), David Audretsch and Mark Sanders give pride of place to multinational enterprises and the process of off-shoring. They develop a global life cycle model of relocation of production to developing countries. This model is based on the allocation of entrepreneurial talent in the North over three alternatives: product innovation, process innovation, and outsourcing/off-shoring to the South. The system is influenced by three major external shocks: (i) the reduction of risks in international transactions due to the collapse of communism, (ii) the entry of populous developing countries into international trade, and (iii) the introduction of general purpose ICT technologies.

The model distinguishes between three stages in the product life cycle. In the first stage new products are discovered. In the second stage, the products mature and innovation focuses on process innovations which reduce costs. As the production process is standardized and cost considerations come to predominate, developing countries achieve a comparative advantage through low labour costs. In the third stage, entrepreneurial decision-making focuses on outsourcing low-cost production to the developing countries, through FDI. The contribution of the modelling exercise is that it endogenizes the allocation of entrepreneurship between the three alternatives: product innovation, process innovation, and outsourcing.

Globalization and technological innovation drive the structural shift towards what the authors call an entrepreneurial economy in the advanced economies in the North. In the entrepreneurial economy, the entrepreneurs of small and medium-sized firms play a key role in employment creation and innovation. Production focuses on innovative high value added activities. There is relative deindustrialization and a shift in labour demand towards skilled non-production workers and off-shoring by Northern entrepreneurs fuels Southern industrialization, based on low-cost labour.

In the context of this volume, one comment is in order. Developing countries are modelled as passive recipients of outsourcing and FDI decisions by entrepreneurs in the advanced economies. The model underplays the importance of absorptive capacities and innovative technology adoption in the developing country context. But, the model provides a powerful summary of past trends. Arguing that innovation primarily takes place in the advanced economies, it describes the impact of innovation on developing countries as an outcome of entrepreneurial choices in the advanced countries.

In Chapter 3 'Firm Ownership and Entrepreneurship', Alice Amsden emphasizes the key entrepreneurial role played in catch-up countries by large privately owned domestic enterprises and conglomerates (POEs). She argues that POEs are much more dynamic than foreign-owned enterprises (FOEs) and have a more substantial impact on development than FOEs, which operate more bureaucratically and are less innovative. Her analysis stands in contrast to the model of Audretsch and Sanders which emphasizes the role of FDI in international technology transfer. Amsden makes the case that the more important impacts of innovation in developing countries derive from indigenous, privately owned firms rather than multinational firms. Her chapter also serves as a counterweight for several chapters in this book that focus on the dynamic role of small and medium-sized firms (see for example the chapters by Stam and van Stel (Chapter 4), Gebreeyesus (Chapter 6), Bascavusoglu-Moreau (Chapter 7), and Voeten, de Haan, and de Groot (Chapter 5)).

Amsden places the dynamic role of POEs in the context of the history of decolonization—hence her chapter has an institutional perspective which complements that of the chapters in Part III of the book. She claims that the successful catch-up countries are the ones which not only got rid of their foreign rulers but also of their foreign firms. After the Second World War, many important industries in developing countries were developed by state-owned enterprises, some of which were privatized at some later stage. In Latin America and Asia, state-owned enterprises were instrumental in starting or advancing the petrochemical and steel industries. In the global privatization drive many state-owned enterprises were privatized. Some were sold to foreign-owned enterprises. But some of the best state-owned enterprises became private-owned enterprises, such as Embraer and Sunkyong.

FOEs have been more important in Latin America than in East Asia. Amsden argues that this may well help explain Latin America's relatively weak performance after the Second World War.

She makes the important point that the manufacturing experience that a dozen developing countries derived from pre-war foreign and national enterprises was the key to post-war economic development. According to Amsden, no developing country entered the orbit of modern world industry after the Second World War, with its own POEs, that did not have pre-war manufacturing experience (see also Amsden 2001; Szirmai 2011). No developing country cultivated a strong cadre of nationally owned firms from scratch. This observation recalls Arthur Lewis's early remark that it takes at least 30–40 years to learn to industrialize (Lewis 1978).

The main contrast in the chapter is between foreign subsidiaries of multinational firms and private national companies. Amsden's characterization of the subsidiaries of multinationals as bureaucratic and inflexible contrasts with recent research emphasizing the increasing autonomy of subsidiaries of global multinationals and the increasing decentralization of knowledge networks (Cantwell 2008). It disregards the present relocation of international R&D facilities to the emerging economies. Also, the recent literature emphasizes that the road taken by Korea and Japan in the past is no longer open to developing countries. Economies which were previously characterized by an excessive reliance on state-owned firms or domestically owned private firms are rapidly opening up to foreign investment and access to global technology seems to be more dependent on FDI than in the past (Lall and Narula 2006).

The debate about the merits of different types of entrepreneurship deserves more attention in future research. A tentative conclusion one might derive from this interesting debate is that absorptive capacity and capacity for upgrading depends on some kind of appropriate balance between privately owned and foreign-owned enterprises. Where foreign-owned firms predominate excessively and large privately owned indigenous enterprises and entrepreneurs are weak or absent, the country may be hampered in its technological and economic development.

Chapter 4 by Erik Stam and André van Stel, entitled 'Types of Entrepreneurship and Economic Growth', continues with the distinction between advanced and developing countries made by Audretsch and Sanders. This chapter is an empirical exploration of the types of entrepreneurship and its impact on economic growth in developing and transition countries. It relates indicators of entrepreneurship to average rates of economic growth in the period 2002–5. For this the authors utilize a dataset on entrepreneurship in 36 countries from the GEM, collected in 2002. Their chapter can be seen as an extension of an earlier contribution of Wong, Ho, and Autio (2005), who—also using GEM data—established that employment creation in advanced economies is driven by a sub-set of 'fast-growing' firms, rather than by start-ups in general.

In their theoretical introduction Stam and van Stel follow Audretsch and Keilbach (2004) in including entrepreneurship as one of the proximate sources of growth in the production function, along with labour, capital, and knowledge/technology. Entrepreneurship is seen as the factor that creates wealth by combining existing production factors in new ways. However, entrepreneurship will only unlock economic development if a proper institutional setting is in place.

In line with our introductory comments, Stam and van Stel emphasize that the role of entrepreneurs in developing countries differs from that in the advanced economies. Their main role in earlier stages of economic development is to discover whether goods, well established in world markets, can be produced at home at low-cost.

Stam and van Stel hypothesize that small owner-operated firms will be the prime movers in the process of structural change in developing countries and transition economies. Their reasoning is that in many developing countries large firms are underrepresented, while in transition economies large organizations are caught up in a process of restructuring. These interesting hypotheses are not supported by direct empirical evidence, but they are used to structure the subsequent analysis in the chapter. The fact that large firms are underrepresented seems to contradict the remarks made by Amsden about the role of large private-owned companies. But, in part this may well depend on the kind of countries one is examining.

In the econometric analysis, the authors use the following indicators for the existence of dynamic entrepreneurship: the Young Business (YB) index in 2002—this is the percentage of the adult population who is the owner/manager of a business less than 42 months old and expects to employ 20 employees within five years (YB high-growth expectation), respectively six employees within five years (YB medium-growth expectation). The dependent variable is the growth of GDP between 2002 and 2005. The authors take institutional characteristics into account by distinguishing three groups of countries: high-income economies, transition economies, and low-income developing economies.

They find that indicators of YB activity have a significant impact on growth in the high-income countries and the transition countries, but not in developing countries. In transition countries growth-oriented entrepreneurs make an important contribution to economic growth. The authors explain the lack of significant effects in developing countries by pointing to the lack of complementary physical and human capital and the scarcity of larger companies that can act as a training ground for SMEs. Their findings imply that the kind of social capabilities—or absorptive capacities—at national level emphasized by Abramovitz (1989) are lacking in developing countries. Also much entrepreneurship in developing countries is survival entrepreneurship which does not make much contribution to self-discovery and structural change. One could also think of a variety of institutional constraints which neutralize the effects of entrepreneurship. This issue is taken up in the chapters on institutions and policies in Part III of the book.

The chapters in this book raise interesting questions about the roles of different types of firms and associated types of entrepreneurship in developing countries: large private-owned firms, state-owned firms, foreign-owned subsidiaries of multinational companies, small and medium-sized enterprises or microenterprises. Amsden points to the key role of large privately owned-firms and conglomerates in Korea and other East Asian economies. This view also finds some support in the chapters by Mani, Athreye, or Toivanen. In Chapter 9 Mani discusses the emergence of knowledge-intensive entrepreneurship in India, which contributes to the acceleration of growth performance after 1991. Athreye (Chapter 11) analyses the business strategies of the dynamic private firms, which transformed the Indian IT sector and the Indian economy as a whole. In a very different setting, Toivanen (Chapter 8) emphasizes the defining role of very large high-tech companies in an advanced economy like Finland.

Other chapters make a case for the importance of small-scale and medium-sized enterprises in the process of economic development. Besides Stam and van Stel, this includes Voeten, de Haan, and de Groot, Bascavusoglu-Moreau, Gebreeyesus, and Akçomak. This is an important line of inquiry, as small firms predominate in the poorest developing countries and are also increasingly important in the advanced economies.

Audretsch and Sanders document the important role of small firms in the USA and other advanced economies, where their shares in employment range from 30 to 72 per cent (see also Acs and Audretsch 1993). Most new jobs emanate from small firms. Small firms account for a substantial share of value added; though this share is always lower than their share in the labour force, as small firms are less capital-intensive than large ones.

Voeten, de Haan, and de Groot show that alongside the large state-owned enterprises in Vietnam, there is a large and vibrant medium and small-scale sector which also engages in innovation. In Vietnam, small firms account for 99.9 per cent of establishments, 77.3 per cent of employment and approximately 26 per cent of GDP in Vietnam. Similarly, Bascavusoglu-Moreau emphasizes the importance of SMEs in Turkey. Turkish growth in the last decade relied heavily on the SMEs, whose dynamism derives from profitability and flexible labour markets. Small firms contributed 61.1 per cent to total Turkish employment around 2002 and 27.3 per cent to GDP. Gebreeyesus shows that microenterprises are the major source of employment and income for the urban population in Africa. He cites ILO statistics that indicate that microenterprises in the informal sector account for half or more of non-agricultural employment in the developing world and 72 per cent in Sub-Saharan Africa (ILO 2002). The same source suggests that these enterprises account for 41 per cent of non-agricultural GDP in Africa.

If we accept that small-scale enterprises are important in many developing countries, in particular the poorest ones, the question arises how innovative small-scale entrepreneurs can be. One should realize that most microenterprises are run by survival entrepreneurs, who may not be in the position to be very innovative or dynamic. Three chapters (Voeten, de Haan, and de Groot, Bascavusoglu-Moreau, and Gebreeyesus) explicitly zoom in on the question how innovative small firms are and what determines why some entrepreneurs are more innovative than others.

In Chapter 5, 'Can Small Firms Innovate? The Case of Clusters of Small Producers in Northern Vietnam', Jaap Voeten, Job de Haan, and Gerard de Groot present a strong case for the existence and importance of innovation in small and medium-sized enterprises in Vietnam in the wider context of the Vietnamese economic reform programme. Their chapter has two aims: methodological and empirical. The methodological part discusses innovation concepts, their relevance to SMEs in developing countries, and the ways in which they can be operationalized in qualitative research. The authors discuss three criteria for the identification of innovation: newness, value creation, and process (which refer to interactions and learning). For each of these criteria, the authors develop three operationalizations. The criteria for newness are the occurrence of product and process innovations, observable differences with past practices introduced by one actor and spreading to others and the subjective perception that there is a breakthrough. The criteria for value creation are lower input costs, higher sales volumes, market expansion, and entry into new markets. The process criteria are somewhat less well-developed than the others. They include feedback and interactions. One could argue that interactions in a system of innovation are among the conditions stimulating innovativeness, rather than being part of its definition.

Empirically the chapter considers handicraft clusters in four Vietnamese villages, which have introduced innovations that are new to the firm, allowing the firms to upgrade production and enter export markets. The innovations were fairly simple ones, such as the introduction of a new gas oven for ceramics, the introduction of new products like children's sweets by cassava producers, the establishment of new retail shops and broader product ranges in silk weaving and tailoring, and the development of new processed bamboo products sold directly to IKEA.

The authors apply their criteria for innovation to four rich qualitative case studies and conclude that real innovation takes place in three of the four cases. The only case which does not meet the criteria for innovation is that of the processed bamboo cluster. Here the innovation was introduced from outside by an NGO and the producers continued to remain dependent on one single buyer, IKEA. In the other cases, the innovation originated from within the cluster. There were many interactions with the outside world.

Innovative success was achieved in spite of the lack of systematic support or technology transfer from outside the cluster. The authors conclude that the Vietnamese system of innovation is still rather underdeveloped. It did not provide much support to the entrepreneurs in the cluster.

The chapter provides persuasive evidence for the existence of innovation capacity amongst small-scale artisanal producers in Vietnam. The authors also conclude that their innovation identification tool was effective in discriminating between innovating and non-innovating clusters.

The chapter defines entrepreneurship in a functional manner, but in fact it focuses exclusively on owner-operated small-scale firms. The innovations are simple innovations that are new to the firm, but which do contribute substantially to firm and cluster dynamism. The question does remain whether the success of the *doi moi* reform process in Vietnam depends on these small enterprises which only account for a quarter of GDP. But it is clear that these small firms are part of the larger process of dynamic transformation of the Vietnamese economy.

The next chapter by Mulu Gebreeyesus, 'Innovation and Microenterprise Growth in Ethiopia', focuses on microenterprises in the informal urban sector of Ethiopia. He analyses a dataset of 974 urban enterprises with less than ten persons engaged, collected in six major Ethiopian cities.

His concept of innovation is comparable to that of Voeten, de Haan, and de Groot. Everything that is new to the firm in the given setting is considered an innovation. This can include quality improvements, better design, increased variety of products, but also the installation of additional machinery, improvements of the premises, changes in accounting or marketing methods, or the hiring of skilled workers. Gebreeyesus argues that in the Ethiopian informal sector innovation is incremental rather than radical. It involves the adoption of product technologies, process technologies, or methods that have already been adopted elsewhere. They are new to the firm, but not to the world or the region, not even to the country or the industry (van Dijk and Sandee 2002). In total 333 firms (34 per cent of the sample) reported one of these kinds of innovation.

This chapter raises two main questions: what are the determinants of innovative performance and do innovative firms show more employment growth than non-innovative firms? These questions are tackled in a cross-section regression framework.

With regard to the first question, Gebreeyesus concludes that innovative performance is significantly related to a number of attributes of entrepreneurs and firms. Younger entrepreneurs are more likely to innovate than older ones, male entrepreneurs more than female entrepreneurs. Among the human capital variables only vocational training has a significant and positive effect on innovation. With regard to firm characteristics, larger firms tend to innovate more than smaller ones. Older enterprises tend to innovate more than younger ones. This last finding somewhat contradicts the result that younger entrepreneurs innovate more. One would expect younger entrepreneurs to be associated with younger firms. However, the relationship between firm age and innovation is curvilinear. Beyond a certain age firms indeed become less innovative.

The most important finding with regard to employment (measured as the cumulative change in employment from 2001–3) is that innovativeness is positively related to growth. Innovative enterprises create significantly more job growth than other enterprises. This positive correlation between innovative performance and growth is observed, controlling for other relevant variables found in the literature, including credit constraints, firm age and size, and entrepreneur characteristics.

This chapter gives us valuable quantitative information and insights about entrepreneurs in the informal sector in an African low-income economy. However, it does not answer the question whether the microenterprises in the informal sector are a potential source of dynamism and growth in the economy at large. In Vietnam, Voeten, de Haan, and de Groot showed that the small-scale sector indeed plays this positive role. But other research on African entrepreneurship (e.g. Rooks, Szirmai, and Sserwanga 2010) indicates that most of the small-scale entrepreneurs are indeed primarily survival entrepreneurs who have little to contribute to the acceleration of growth in the wider economy. This question of the potential contributions of microenterprises to innovation and national dynamics is an important one, which should remain high on the research agenda.

In the last chapter in Part II, Chapter 7, entitled 'Entrepreneurship and the National System of Innovation: What is Missing in Turkey?' Elif Bascavusoglu-Moreau focuses on the national system of innovation as an important determinant of innovative performance of small and medium-sized enterprises in Turkey using data from a large-scale survey held amongst 50,000 Turkish SMEs in 2005. She finds very low levels of innovative activities amongst Turkish small firms. In the period 2004–6 only 31.4 per cent of enterprises were active in implementing innovations. The smaller the enterprise, the less innovative it seems to be. Bascavusoglu-Moreau investigates whether this weak innovative performance is due to a lack of entrepreneurial capabilities or due to shortcomings in Turkey's national system of innovation.

Her analysis shows that Turkish growth rates in the last decade depend heavily on the performance of small firms. These firms can be drivers of growth, even without being very innovative. This implies that their contribution to growth would be even greater if they could become more innovative.

The chapter is based on a regression analysis, estimating a knowledge production function, with dummies for patents, trademarks, and utility models as dependent variables, and a mix of firm-level and district-level variables as explanatory variables. Firm-level determinants of innovative performance include investment in R&D, use of ICT, export activities, outsourcing, educational level of employees, net current assets, and the firms' ability to learn, such as certified management systems and/or quality labels.

Bascavusoglu-Moreau argues that the size of the shadow economy in Turkey is one of the obstacles to innovativeness. Though informality provides advantages such as low-cost and flexibility, it limits firms' access to capital markets and traps potentially innovative entrepreneurs in low-productivity sectors. The choice for operating in the informal sector is encouraged by the heavy tax burdens on small firms.

The Turkish national innovation system proves rather weak at fostering the innovative performance of SMEs. Contrary to expectations, there is no stimulus to innovation coming from being located in industrial estates, nor from the size of these industrial estates. Furthermore, weak innovative performance seems due to financial constraints and a lack of market opportunities and support and incentive systems. One of the most pronounced weaknesses of the Turkish innovation system is the lack of cooperation and linkages between different actors. The elements of the innovation system are in place, but they do not interact sufficiently. The author reflects on policy implications to deal with these systemic weaknesses.

1.4.2 *Policies and institutions for innovation in developing countries*

In section 1.3.3, we highlighted that the extent of the market and appropriate innovation policies are important for innovation, given that innovation requires scale, specialization, and the exploitation of externalities. The extent of the market and the nature of innovation policies are often the result of the 'reward structure', or the institutional framework, found in a particular country. Good policies, and appropriate market incentives require good institutions. What are good policies and institutions for innovation in developing countries?

The question is so central to current debates on development and entrepreneurship, that all of the chapters in this book have some reference to policies and institutions. For instance, Audretsch and Sanders suggest that global developments and the closer integration of North and South through globalization will advance development in the South through technological innovation in the North. Amsden shows that historical legacies and national leadership have lasting impacts on the innovative behaviour of a country's entrepreneurs. Stam and van Stel explicitly recognize that the quality of entrepreneurship (and its impact) will differ across different levels of institutions (proxied by development outcomes).

While all of the previous chapters have implications for policies and institutions, the chapters in part III of the book explicitly focus on these issues. The chapter by Bascavusoglu-Moreau (Chapter 7) provides a pivotal point in the book. Here, the emphasis starts to shift from the primary analysis of the impacts and determinants of innovation to the question how policies and institutions affect innovation in developing countries. Thus, she concludes from her analysis that the innovative weaknesses of Turkish small firms are more related to shortcomings in regional systems of innovation than a lack of entrepreneurial capabilities at firm-level. In many developing countries and emerging economies, there is an ample supply of entrepreneurship. Inadequate innovation performance therefore primarily reflects a lack of sound policies and appropriate institutions.

In Chapter 8 'Innovation Policy, Entrepreneurship, and Development: A Finnish View', Otto Toivanen scrutinizes one of the most frequently applied instruments of innovation policy, namely R&D subsidies. He does this within a particular institutional setting that is the Finnish innovation support system. Although Finland is an advanced economy, it initially lagged behind the rest of Europe and then managed to converge rapidly. Furthermore, it has done so in particular through high-technology industrialization. As such, its experience is very relevant for emerging economies.

Toivanen argues that there are sound theoretical justifications for governments to provide R&D subsidies. This is due to the positive externalities that result from innovation—these include consumer surpluses and knowledge spillovers—as well as financial market imperfections that constrain innovative entrepreneurs. R&D subsidies address the problem of heterogeneity. In Toivanen's words 'firms, R&D projects, and innovations are highly heterogeneous . . . a policy that is optimal . . . should vary not only from firm to firm, but from project to project'.

Despite a strong theoretical case, Toivanen is at pains to point out that the complexity of crafting an appropriate R&D policy is significant, particularly given the informational challenges facing governments. Furthermore, while R&D subsidies can be useful to realize externalities and can be tailor-made to fit an entrepreneur's needs, they also have a number of shortcomings as a tool to promote innovation in small enterprises. First, the application procedures are costly. Next, R&D subsidies are often paid out ex post, which means that they are not ideal for cash strapped firms, which might rather need up-front financing for innovation. In such cases, tax credits might be a viable alternative. Also, the process of tailoring R&D subsidies takes considerable time, during which firms may lose any potential first mover advantage from their innovations.

Toivanen concludes by pointing out that while the Finnish R&D system is well managed, it is still not clear how much this system contributed to the country's economic success. This is not to say that the well-run Finnish innovation system did not contribute towards the country's growth and development—it is a very likely candidate—(see also Autio and Parhankangas 1998; Autio and Yli-Renko 1998). However, its precise contribution needs to be understood within the broader context of Finland's post-Second World War development strategy, which, in addition to the promotion of innovation, also included heavy government intervention in terms of investment, credit rationing, wage moderation, as well as repeated devaluations. Readers interested in Finland's post-war catch-up are referred to Jäntti and Vartiainen (2009).

Toivanen warns that it is not clear to what extent the Finnish R&D system can be copied by developing countries. It requires well-developed institutional and administrative capabilities, which are not available in all countries. He stresses that any effective innovation support system will require prior investments in education, labour skills, and human capital. Thus, the Finnish experience will probably be more relevant to emerging economies in Latin America, than developing countries in Sub-Saharan Africa.

The next chapter (Chapter 9) by Sunil Mani is entitled 'Promoting Knowledge-intensive Entrepreneurship in India'. The chapter starts out by demonstrating how important private technology-based entrepreneurial firms have been for the acceleration of economic growth in India in the second half of the 1990s. Mani takes a clear position in the debate on liberalization in India. He argues that the post-1991 liberalization provided key institutional changes, which created new opportunities for dynamic entrepreneurship, in essence by expanding the market. These changes included the abolition of a variety of regulations governing and limiting entry of new firms and expansion, diversification of industrial activity, and acquisition of technology from abroad. Licensing was discontinued, foreign firms were allowed to enter the economy, acquisition of foreign technology became much easier, and reservations for public sector enterprises were abolished.

Mani documents the growth in entrepreneurship and its increasing knowledge-intensity. After 1992, an average of 34,000 new firms were established each year. Especially since 2003/4 an increasing share of these firms was technology-based, for instance in IT software, biotechnology, aerospace, and telecoms.

The remainder of the chapter outlines the policy and institutional environment that gave rise to the growth in knowledge-intensive entrepreneurship in India. Mani identifies five broad facilitating factors: (i) the liberalization of the economy that created many new market opportunities, (ii) the general increase in financial resources for innovation and entrepreneurship, including in particular venture capital, (iii) a rise in government support programmes and public–private partnerships, (iv) the emergence of private institutions and initiatives to complement government support programmes for innovation, and (v) the increased availability of skilled labour, essential for high-technology products and services.

Mani stresses four key policy challenges for the government, namely (i) to create the right environment for doing business and starting new enterprises, (ii) to ensure the formation of skills, (iii) to ensure access to risk capital, and (iv) to facilitate networking and organizational learning. By and large, Mani comes to a positive assessment of Indian policy reform. But as the chapter by Athreye also makes clear, major challenges remain.

One novel government-led initiative was the establishment of business incubators. Mani points out that by 2007 there were already 40 incubators in India—a number that is still relatively small compared to other countries, but of sufficient importance for Mani to describe it as 'an experiment which is worth watching'.

The role of incubators in developing countries comes under detailed scrutiny in Semih Akçomak's chapter 'Incubators as Tools for Entrepreneurship Promotion in Developing Countries'. Akçomak starts off by revisiting some of the determinants of innovation discussed in greater detail in Part II of this book. He provides a useful reminder that the predominantly small firms in developing countries face many size-related disadvantages, and that 'small size is an important constraint for process and product innovations'. In developing countries this constraint is often compounded by poorly functioning formal institutions, a difficult business environment, a lack of support and complementary inputs, including human capital. These constraints to innovation resonate with Mani's analysis of the ways in which India's policy and institutional reforms contributed to the increase in process and product innovations since 1991.

Incubators have been used in a growing number of countries over the past two decades as a tool to help alleviate the constraints on innovation mentioned in the previous paragraph. Incubators differ from related policy instruments such as industrial estates, techno parks, and science parks, in terms of higher technology levels and higher degrees of management support. Incubators are increasingly being adopted by developing countries in order to overcome some of the weaknesses in their institutional environment. Akçomak provides an overview of the extant literature on incubators, as well as a novel typology of incubators. Incubators are classified along two dimensions: profit versus not-for-profit and general versus specific. He also describes their evolution over time. The new generations of incubators developed since 2000 are both for profit and highly specific. In contrast, the oldest generations of incubators set up in the 1960s were general and not-for-profit.

The chapter reviews the evidence of the effects of incubators and discusses country case studies for Brazil, China, India, and Turkey. He concludes that the challenges faced by incubators should not be underestimated. There is no guarantee that an incubator will succeed in fostering innovation and the creation of sustainable new (innovative) firms. To maximize the chances of success he sets out eight dimensions of a good incubator policy. Success factors include (i) clarity of mission and purpose, (ii) clear selection, entry and exit criteria, (iii) managerial capacity and incubator management skills, (iv) regular monitoring and performance evaluation of participating firms, (v) strategic selection of services, minimizing start-up costs, and red tape, (vi) focus on intangible services rather than tangible services such as office space or infrastructure, (vii) promoting networking as a deliberate strategy, and (viii) financial sustainability. Incubators should eventually become financially self-sustaining and less dependent on public funding.

In the final chapter, Chapter 11, 'Overcoming Adversity in Entrepreneurship-led Growth: Evidence from the Indian Software Sector', Suma Athreye analyses the interplay of institutions and business strategies. While Mani emphasized the importance of liberalization and institutional reform since 1991, Athreye highlights the role of the adverse environment and the difficult conditions for doing business in the pre-reform period. Athreye tackles the puzzle of how entrepreneurs succeeded in innovating in an often deeply adverse environment, characterized by overregulation, high cost of doing business, weak enforcement of property rights, poor capital markets, and underdeveloped markets. Remarkably, Indian software firms found a way around all these obstacles, by creatively choosing appropriate business models and capitalizing on abundant and cheap high skilled labour. Adversity promoted creativity. Athreye discusses a wide range of business models, but the main model was the outsourcing model, where Indian software firms provided services for international clients and used this process to develop their own capabilities.

Her second finding is that the very success of the software industry was a source of improvement in the institutional environment. In this respect she is interested in the phenomenon of 'institutional entrepreneurs'. An institutional entrepreneur has been described as 'an innovative person who starts or expands his business venture and in the process helps destroy the prevailing non-market institutions in order for his business to be successful' (Li, Feng, and Jiang 2006: 358). Li, Feng, and Jiang describe the case of institutional entrepreneurship in China. The chapter by Athreye provides a useful complement to their work, dealing with India where the initial institutional environment and policy challenges were very different. Indeed for Athreye there is little doubt that, just as institutional and policy improvements stimulated China's impressive growth, this was also the case in India after 1991. In her words 'the spectacular growth of industry in the 1990s was also marked by an improvement in the institutional infrastructure surrounding the software outsourcing industry, which generally served to ease constraints on the industry's further growth. These included capital and labour market reform, better access to finance, improved IP protection and contract enforcement'. The crux of the chapter is that these changes in the policy and institutional environment for innovation were brought about not by the government taking the lead, but through institutional entrepreneurship. As she puts it, 'the impetus for institutional reform has not come from the government, international institutions or their advisors, but primarily from the business sector itself'. The sectoral organization NASSCOM acted as an effective lobbyist for institutional reform, which then spread to other sectors of the economy.

This chapter highlights the endogenous nature of institutional change as against the view that one first needs a major overhaul of institutions in order to remove obstacles to growth (c.f. Rodrik 2006). But a key question remains: why did the Indian entrepreneurs overcome institutional constraints, while this did not always happen elsewhere? Athreye seeks part of the answer to this question in the sustained period of buoyant export demand for IT services, which allowed Indian entrepreneurs to circumvent domestic constraints.

1.5 Conclusion

Entrepreneurship can contribute in important ways to economic development. One way it does this is through innovation, which involves the development of new products, new processes, new sources of supply, but also the exploitation of new markets and the development of new ways to organize business. However, not all entrepreneurs innovate. A better understanding of the determinants and impacts of innovation in developing countries and the policies and institutions that support or hinder innovation is the central contribution of the chapters of this book. Without detracting from the wealth of detail contained in the various chapters, we conclude this introduction by spelling out three general lessons.

First, the impact of innovation is important across different countries and institutional contexts. But the nature and role of innovation will differ at different levels of economic development. Entrepreneurs in low-income developing countries provide innovations that are important for firm and country growth, even if they are incremental in nature. Innovation in developing countries involves the process by which firms master and implement the design and production of goods and services that are new to them. Many small improvements in product design and quality, changes in the way production is organized, creativity in marketing and modifications in production processes and techniques reduce costs, increase efficiency and flexibility to respond to changes in competitive conditions and enhance productivity and employment growth. In emerging economies innovation involves upgrading and shifting to higher levels of technological sophistication. These findings give lie to the belief that innovation matters only for the most advanced economies. Innovation plays an important role in catch-up and growth in a globalized economy. First and foremost this is due to the varied innovations realized by local entrepreneurs in developing countries. But it also depends on the activities of entrepreneurs in advanced economies, where innovations are generated that subsequently spread to developing countries, to be applied in their particular local contexts.

Second, the conditions under which entrepreneurs innovate in developing countries are very diverse. This book deliberately sets out to cover a wide range of countries, utilizing a variety of quantitative and qualitative research methods, and focusing on different types of firms, ranging from SMEs and informal sector firms to large multinationals. Large and small firms can both be innovative, but in very different ways. In this book special emphasis has been given to small and medium-sized enterprises, since these predominate in many developing economies. They contribute to growth, but not optimally since they also face particular constraints to innovation. How entrepreneurs respond to these challenges depends on the characteristics of the entrepreneurs such as education, age, managerial experience, and access to networks.

Third, the policy and institutional environment is an important determinant of entrepreneurs' innovative behaviour. This explains why entrepreneurs with very similar characteristics and features may differ in their innovative performance. An effective system of innovation and a supportive institutional environment is required to spur innovation. Several of the chapters have pointed to the need for direct and indirect government support for innovation. This can take many forms, including improvements in the environment for doing business, R&D subsidies, provision of risk and venture capital, tapping into the potential of migrant workers and diasporas, expanding technical and managerial education, improving infrastructure, and active public–private partnerships. Sometimes, even an adverse environment can spur innovative behaviour, and entrepreneurs themselves may become the drivers of policy and institutional change, rather than only being determined by their environments.

Entrepreneurs' innovation will be of increasing importance given the current and future challenges facing global development. These include persistent global inequalities, growing vulnerability of countries to external shocks, including financial crises, natural disasters, and the threats of climate change. To meet these challenges—which are also opportunities—innovative entrepreneurship is indispensible. In the light of these challenges, the very concept of entrepreneurship needs to be broadened. In particular, we envisage growing roles for social entrepreneurship, public entrepreneurship, institutional entrepreneurship, and even non-state sovereign entrepreneurship. These topics fall outside the scope of the present book, and constitute an agenda for future research.

Two further areas for future research emerge from this book. First, there is a need to better understand the innovative contributions of different types of firms and different types of entrepreneurship in different country settings. Is innovation driven by dynamic small and medium-sized firms or by large conglomerates? What are the respective contributions of domestic and multinational firms? What role can public enterprises play in innovation? What is the innovative potential of the millions of microenterprises in developing countries? We have learned from the contributions in this book that different types of entrepreneurship predominate in different settings and at different levels of development. But we need to go beyond this, to try to identify patterns which are appropriate for different settings and levels of development.

Second, various authors in this book have pointed to the crucial relationship between institutions and entrepreneurs in the process of innovation. High on the research agenda remains the question why in some cases the institutional environment effectively stifles innovative behaviour, while at other times entrepreneurs can find a way around the obstacles? There may be special windows of opportunity, as in the case of the software sector in India. It would also seem that the overall level of development plays a role here. In emerging economies such as China, India, and Turkey the foundations for the rise of innovative firms have been laid. Firms can develop and prosper in spite of a variety of institutional shortcomings, because there are sufficient business opportunities and incentives. In a low-income setting the institutional constraints may well be overwhelming.

We have learned that entrepreneurs are not only passive victims of obstacles. They can actively shape institutions. How this happens, and how such institutions stimulate further innovation, needs to be further examined. Finally, the appropriate design of institutions, organizations, and policies to promote both entrepreneurship and innovation remains a formidable challenge, complicated not only by the lack of sufficient government capacity and resources in developing countries, but also by a lack of rigorous, evidence-based research. An important task awaits researchers interested in working at the intersection of the fields of entrepreneurship, development economics, and innovation studies. We hope that this book will stimulate further research in these directions.

Notes

1. For instance, Acs and Szerb (2009) and Porter et al. (2004) argue that economic development proceeds through three stages: (i) a factor-driven stage, (ii) an efficiency-driven stage, and (iii) an innovation-driven stage. Hence it is only at advanced levels of development that growth is believed to be 'innovation-driven'. Acs and Szerb (2009: 351) report estimates in which innovation accounts for 'only about 5 per cent of economic activity' in poor countries, compared to 30 per cent in advanced economies.
2. There are many definitions of entrepreneurship, because it is studied from a great variety of disciplinary perspectives. Within economics, however, there is a large degree of concurrence of definitions (Casson 2003). The definition we use here is consistent with most definitions in the fields of economics and management.
3. There is a wide variety of definitions of innovation, which are reflected in the various chapters in this book. The Oxford Handbook of Innovation shies away from a formal definition of innovation, but presents some key characteristics of innovation. Chapter 5 of this book provides a useful overview of the different definitions and concepts.
4. In the development literature on innovation the relevance of traditional secondary indicators is questioned for the poorest developing countries, where firms hardly do any formal R&D nor apply for patents. Instead innovation surveys are increasingly used as a tool for measuring technological and innovation activity in firms. The new-to-the-firm aspect of the innovation concept that is key to the innovation surveys indeed allows studying the distinguishing characteristics and strategies of innovating firms, at any stage of economic development of the country (UNU-INTECH 2005; OECD/EUROSTAT 2005).
5. As documented by Aubert (2004: 6) 'innovation policy' originated in the 1964 'Charpie Report' which was submitted to President Johnson.

References

Abramovitz, M. (1989) 'Thinking about Growth'. In M. Abramovitz (ed.), *Thinking about Growth and other Essays on Economic Growth and Welfare*. Cambridge: Cambridge University Press.

Acs, Z., and D. Audretsch (1993) *Small Firms and Entrepreneurship: An East–West Perspective*. Cambridge: Cambridge University Press.

Acs, Z., and D. Storey (2004) 'Introduction: Entrepreneurship and Economic Development', *Regional Studies*, 38 (8): 871–7.

Acs, Z., and L. Szerb (2009) 'The Global Entrepreneurship Index', *Foundations and Trends in Entrepreneurship*, 5 (5): 341–435.

Amsden, A. H. (2001) *The Rise of 'the Rest': Challenges to the West from Late-Industrialising Economies*. New York: Oxford University Press.

Aubert, J.-E. (2004) 'Promoting Innovation in Developing Countries: A Conceptual Framework', Working Paper. Washington, DC: World Bank Institute.

Audretsch, D. B., and M. Keilbach (2004) 'Entrepreneurship Capital and Economic Performance', *Regional Studies*, 38: 949–59.

Autio, E., and A. Parhankangas (1998) 'Employment Generation Potential of New, Technology-Based Firms during a Recessionary Period: The Case of Finland', *Small Business Economics*, 11: 113–23.

Autio, E. and H. Yli-Renko (1998) 'New Technology-Based firms in Small Open Economies—An Analysis Based on the Finnish Experience', *Research Policy*, 26: 973–87.

Baumol, W. J. (1990) 'Entrepreneurship: Productive, Unproductive and Destructive', *The Journal of Political Economy*, 98 (5): 893–921.

Cantwell, J. (2008) *Blurred Boundaries between Firms, and New Boundaries within (Large Multinational) Firms: The Impact of Decentralised Networks for Innovation*. UNU-MERIT seminar presentation, August.

Casson, M. (2003) *The Entrepreneur: An Economic Theory*. 2nd edn. Cheltenham: Edward Elgar.

Fagerberg, J. (2005) 'Innovation: A Guide to the Literature'. In J. Fagerberg, D. C. Mowery, and R. R. Nelson (eds), *The Oxford Handbook of Innovation*. Oxford: Oxford University Press.

Fagerberg, J., D. C. Mowery, and R. R. Nelson (eds) (2005) *The Oxford Handbook of Innovation*. Oxford: Oxford University Press.

Goedhuys, M., N. Janz, P. Mohnen, and J. Mairesse (2008) 'Micro Evidence on Innovation and Development (MEIDE): An Introduction', *European Journal of Development Research*, 20: 167–71.

Goedhuys, M., and L. Sleuwaegen (2010) 'High-growth Entrepreneurial Firms in Africa: A Quantile Regression Approach', *Small Business Economics Journal*, 34: 31–51.

Goedhuys, M., and M. Srholec (2010) 'Understanding Multilevel Interactions in Economic Development', Working Paper 2010-003. Maastricht: UNU-MERIT.

Gries, T., and W. A. Naudé (2010) 'Entrepreneurship and Structural Economic Transformation', *Small Business Economics Journal*, 34 (1): 13–29.

Hausmann, R., and D. Rodrik (2003) 'Economic Development as Self-Discovery', *Journal of Development Economics*, 72 (2): 603–33.

Helpman, E. (2004) *The Mystery of Economic Growth*. Cambridge, MA: Harvard University Press.

Hobday, M., and F. Perini (2009) 'Latecomer Entrepreneurship: A Policy Perspective'. In M. Cimoli, G. Dosi, and J. E. Stiglitz (eds), *Industrial Policy and Development. The Political Economy of Capabilities Accumulation*. Oxford: Oxford University Press.

International Labour Office (ILO) (2002) *Women and Men in the Informal Economy: A Statistical Picture*. Geneva: ILO

Jäntti, M., and J. Vartiainen (2009) 'The Finnish Development State and its Growth Regime', Research Paper 2009/35. UNU-WIDER: Helsinki.

Lall, S., and R. Narula (2006) 'Foreign Direct Investment and its Role in Economic Development. Do We Need a New Agenda?'. In R. Narula and S. Lall (eds), *Understanding FDI-Assisted Economic Development*. Milton Park: Routledge.

Lazonick, W. (2008) 'Entrepreneurial Ventures and the Developmental State: Lessons from the Advanced Economies', Discussion Paper 2008/01. UNU-WIDER: Helsinki.

Leibenstein, H. (1968) 'Entrepreneurship and Development', *American Economic Review*, 58 (2): 72–83.

Lewis, W. A. (1954) 'Economic Development with Unlimited Supplies of Labour', *The Manchester School of Economic and Social Studies*, 22: 139–91

Lewis, W. A. (1978) *The Evolution of the International Order*. Princeton, NJ: Princeton University Press.

Li, D. D., J. Feng, and H. Jiang (2006) 'Institutional Entrepreneurs', *American Economic Review*, 96 (2): 358–62.

Lipsey, R. G., K. I. Carlaw, and C. T. Bekar (2005) *Economic Transformations: General Purpose Technologies and Long-Term Economic Growth*. Oxford: Oxford University Press.

Lundvall, B. A. (1992) *National Systems of Innovation: Towards a Theory of Innovation and Interactive Learning*. London: Pinter.

Minniti, M., and W. A. Naudé (2010) 'What Do We Know About the Patterns and Determinants of Female Entrepreneurship Across Countries?', *The European Journal of Development Research*, 22 (2): 1–17.

Mytelka, L. K. (ed.) (1999) *Competition, Innovation and Competitiveness in Developing Countries*. Paris: OECD.

Naudé, W. A. (2010a) 'Entrepreneurship, Developing Countries and Development Economics: New Approaches and Insights', *Small Business Economics Journal*, 34 (1): 1–12.

Naudé, W. A. (ed.) (2010b) *Entrepreneurship and Economic Development*. Basingstoke: Palgrave Macmillan.

Naudé, W. A. (2010c) 'Entrepreneurship, Global Development and the Policy Challenge', *Harvard College Economic Review*, 4 (2), forthcoming.

Nelson, R. (ed.) (1993). *National Innovation Systems. A Comparative Analysis*. Oxford: Oxford University Press.

OECD/EUROSTAT (2005) *Oslo Manual, Guidelines for Collecting and Interpreting Innovation Data*, 3rd edn. Paris: OECD.

Porter, M. E., K. Schwab, X. Sala-i-Martin, and A. Lopez-Claros (eds) (2004) *The Global Competitiveness Report 2003–2004*. World Economic Forum. New York: Oxford University Press.

Rodrik, D. (2006) 'Goodbye Washington Consensus, Hello Washington Confusion? A Review of the World Bank's Economic Growth in the 1990s: Learning from a Decade of Reform', *Journal of Economic Literature*, XLIV, December: 973–87.

Rogers, E. M. (2003) *Diffusion of Innovations*, 5th edn. New York: The Free Press.

Romer, P. (1990) 'Endogenous Technological Change', *Journal of Political Economy*, 98: S71–S102.

Rooks, G., A. Szirmai, and A. Sserwanga (2010) 'The Interplay of Human and Social Capital in Entrepreneurship in Developing Countries. The Case of Uganda'. In W. A. Naudé (ed.), *Entrepreneurship and Economic Development*. Basingstoke: Palgrave Macmillan.

Schumpeter, J. A. (1934) *The Theory of Economic Development*. Cambridge, MA: Harvard University Press.

Schumpeter, J. A. (1943) *Capitalism, Socialism and Democracy*. London: Allen and Unwin.

Shane, S. and S. Venkataraman (2000) 'The Promise of Entrepreneurship as a Field of Research', *Academy of Management Review*, 25 (1): 217–26.

Szirmai, A. (2005) *The Dynamics of Socio-economic Development: An Introduction*. Cambridge: Cambridge University Press.

Szirmai, A. (2008) 'Explaining Success and Failure in Development', Working Paper 2008–013. Maastricht: UNU-MERIT.

Szirmai, A. (2011) 'Explaining Success and Failure in Economic Development'. In B. van Ark and D. S. Prasada Rao (eds), *World Economic Performance. Past, Present and Future*. Cheltenham: Edward Elgar.

UNU-INTECH (2005) 'Measuring Innovation: Making Innovation Surveys work for Developing Countries', *Technology Policy Brief*, 4 (1). Maastricht: UNU-MERIT.

Van Dijk, M. P., and H. Sandee (2002) 'Innovation and Small Enterprise Development in Developing Countries'. In M. P. van Dijk and H. Sandee (eds), *Innovation and Small Enterprises in the Third World*. Cheltenham: Edward Elgar.

Verspagen, B. (2005) 'Innovation and Economic Growth'. In J. Fagerberg, D. C. Mowery, and R. R. Nelson (eds), *The Oxford Handbook of Innovation*. Oxford: Oxford University Press.

Wong, P. K., Y. P. Ho, and E. Autio (2005) 'Entrepreneurship, Innovation and Economic Growth: Evidence from GEM Data', *Small Business Economics*, 24 (3): 335–50.

Part II

Determinants and Impacts of Innovative Entrepreneurship in Developing Countries

2

Technological Innovation, Entrepreneurship, and Development

David B. Audretsch and Mark Sanders

2.1 Introduction

Globalization and technical change have been identified as key sources of structural change in industrialized and developing countries alike (e.g. Grossman and Helpman 1991; Helpman 1998). At the same time both trends can be understood as the outcome of economic processes rather than exogenous to the global economic system. This begs the question of what fundamental economic mechanisms underlie the interaction between globalization, technical innovation, and structural change. The purpose of this chapter is to address that question. We do so by analysing the impact of three important shocks to the global economy that have occurred since the 1980s.

Arguably, the fall of communism in essence caused an exogenous global reduction in political risks and allowed large low-skilled labour abundant economies to join the global economy (Audretsch 2007). This opening-up of opportunities for economic integration coincided with the, also exogenous, introduction of a new General Purpose Technology (GPT) in the form of Information and Communication Technologies (ICT) over the past 25 years (Helpman 1998).

We argue that this combination of shocks caused the developed economies to make the transition from an industrial to an entrepreneurial economy in response to new opportunities; moreover, the entrepreneurs in the developed countries changed the pattern of national and international specialization in such a way that the developed countries experienced wage divergence among skill groups and increased the share of new, knowledge-intensive products and services in their economy. This corresponds with transition to the 'knowledge economy' and the 'rise of the creative class' that has been well established in the literature (Florida 2003).

On the other hand, through outsourcing and FDI the developing and emerging economies increasingly engage in the production of mature industrial goods (manufactures) and compete on the basis of their (static) comparative advantage, which is low wages. These countries enter the industrial era, but based on technology transfer that is largely initiated by international entrepreneurs.

It is quite clear from the empirical literature (e.g. Keller 1996) that not all countries make that transition equally successfully and that for example absorptive capacity is a key factor in attracting and benefiting from the existing international entrepreneurship. In addition, local entrepreneurs and firms in developing and emerging economies also seize new opportunities and engage in export (one might call that 'insourcing') or even initiate FDI themselves. Moreover, as countries develop, they can be expected to move from agricultural self-sufficiency with abundant low-skilled labour, to industrial export-oriented with increasing levels of human capital to an entrepreneurial and service-oriented mature economy with skilled labour and sophisticated demand. That richness, however, is too much to deal with in one model and we choose to limit ourselves to analysing the world as if it consisted of only two regions with two types of products and corresponding production technologies: one advanced region that produces advanced early stage innovative products using skilled labour and one developing region that produces routine, mature standardized products using less skilled labour. We like to stress at this point that this simplification is not intended to be a statement about the world as it is but rather is a useful abstraction that allows us to zoom in on the mechanism we are primarily interested in.

To outline our basic argument we build on elements from life cycle trade models as in Krugman (1979) and Grossman and Helpman (1991), models of endogenous skill-biased technical change as in Acemoglu (2002b) and Sanders (2005), and notions from the large literature on the entrepreneurial function surveyed by, for example, Audretsch and Acs (2005).

Our model shows how globalization and technological innovation drive the structural shift towards the entrepreneurial economy in the advanced region, but entrepreneurs in turn may drive the process of globalization and GPT exploration, both in the advanced and developing economy. We conclude from our analysis that the opening-up and exploitation of opportunities for trade and GPT-related products and services may have especially hurt the low-skilled workers in advanced economies, but globalization through reduced political risks in the emerging countries unambiguously benefited all workers in both regions.

2.2 Stylized facts

2.2.1 *Political risk, global labour supply, and a new general purpose technology*

Globalization and the advent of new ICT technology can not reasonably be considered in isolation. It is obvious that the spectacular drop in the price of communication has been instrumental in allowing firms to expand their operations globally. It is also more than reasonable to assume that the development of ICT technology has benefited greatly from the boom in international demand for goods and services in general.

Globalization, however, would not have occurred to the degree that it has if the fundamental changes were restricted to the advent of the microprocessor and telecommunications. During the post-war era most trade and economic investment was confined to Europe and North America, and later a few of the Asian countries, principally Japan and the Asian Tigers. Technology alone could not have overthrown that, largely political, constellation. Trade with countries behind the iron curtain was restricted and in some cases prohibited. Even trade with Japan and other Asian and South American countries was highly regulated. Similarly, investments in politically unstable countries resulted in episodes of national takeovers and confiscation where the foreign investors lost all or part of their investments.

We therefore argue that it took a political revolution in significant parts of the world to reap the full benefits of these technological changes. In the final decade of the twentieth century communism has been abandoned as unworkable (China), has imploded (the USSR), or has been overthrown (Central and Eastern Europe, CEE). And the 'victory' of democracy and capitalism over communism also provoked a renewed commitment to the principles of free trade and international competition in countries such as India and Brazil. India, for example, became accessible as a trading and investment partner after opening its economy in the early 1990s. With some of these areas (re)-joining the world economy for the first time in decades, the post-war equilibrium came to a sudden end.

What most distinguished the new players in the global economy was their relative abundance in cheap labour with comparatively low levels of education and productivity. For example, in the early part of the 1990s, the daily earnings of labour were estimated to be US$92.24 in the United States and US $78.34 in the European Union. This stood in sharp contrast to wages of only US$6.14 in Poland and US$6.45 in the Czech Republic shortly after the Berlin Wall fell. With Asia, the wage gap was even greater, as daily earnings were US$1.53 in China, US$2.46 in India and US$1.25 in Sri Lanka where previously trade barriers and political controls kept these economies disconnected

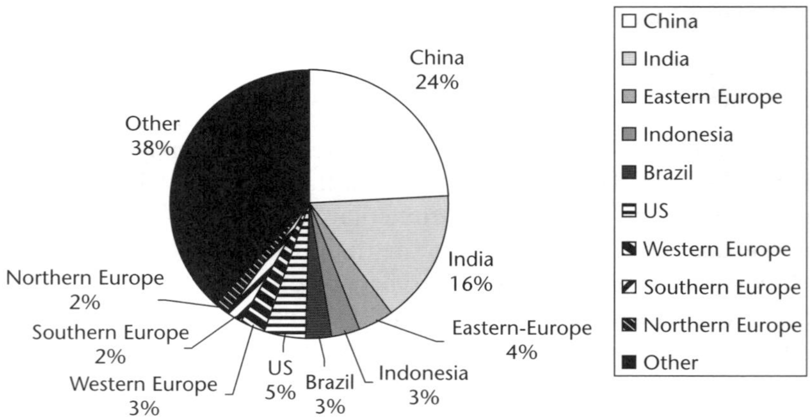

Figure 2.1 Share in global active population in 2005

Source: ILO (2005) at <http://www.ilo.org/public/english/bureau/stat/portal/index.html>.

and allowed such disparities to persist. As Figure 2.1 shows, China and India alone now account for over one third of the world's active population. We thus establish

> Fact 1: *Globalization is characterized by the inclusion of large populous low-wage economies.*

The global economic system thus had to deal with a sudden expansion of the supply of cheap manual labour. In any market economy the response to such a shock is predictable. In the incumbent players' economies the wages decline for the jobs such workers can do and rise in complementary occupations (including the local production of non-tradables). In addition, following standard Heckscher–Ohlin trade theory, one expects a shift in international trade patterns and increasing standards of living in the new entrant countries. The data over the past ten years show all these trends.

In addition the political revolution created opportunities for entrepreneurs associated with the labour cost differences and emerging markets for their products. Trade with foreign companies and especially foreign direct investment was simply prohibited by local governments or considered to be too risky. But with opening up to trade and FDI, formerly closed countries allowed and often invited Northern entrepreneurs in to arbitrage away such differentials by outsourcing and off-shoring, making good money in the process.[1]

As can be verified in Figures 2.2 and 2.3, volumes of FDI increased rapidly between 1990 and 2000. Meanwhile one can see in Figure 2.4 that the rate of return and the yield on foreign assets fell between 1983 and 2001 (by 8 and 3 basis points on average per year, respectively), indicating a drop in the risk

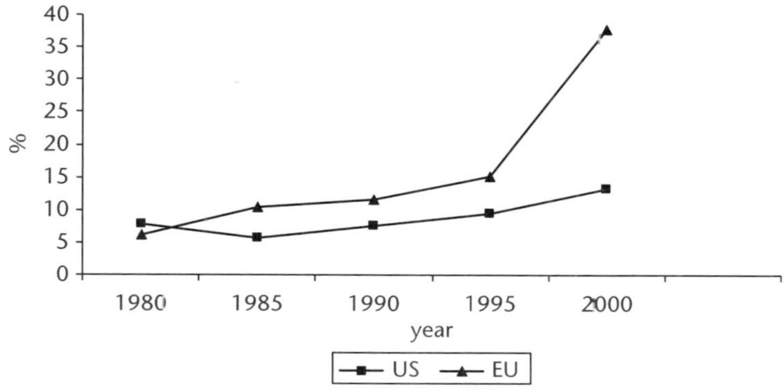

Figure 2.2 FDI outflow as percentage of GDP
Source: UNCTAD (2003).

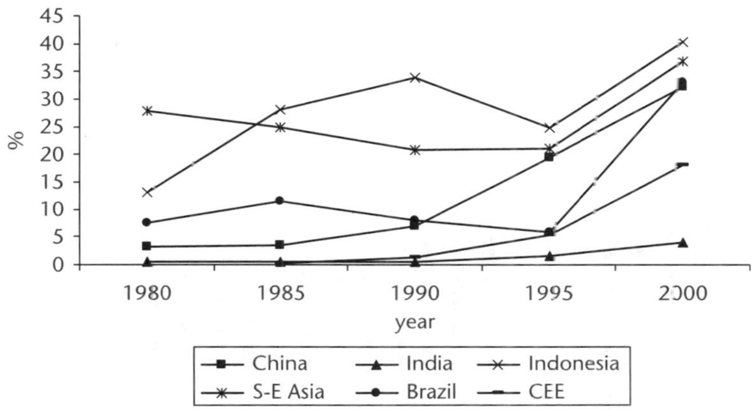

Figure 2.3 FDI inflow as percentage of GDP
Source: UNCTAD (2003).

premium as Western investors apparently were willing to invest larger amounts at lower rates of return.

From these facts, we establish stylized

Fact 2: *Political risks declined worldwide.*

The Northern entrepreneurs were aided and facilitated in setting up such operations by the recently developed ICTs that virtually destroyed distance and provoked serious reorganization of production in itself. This new GPT predates the fall of the Wall, but the North was still in the middle of exploring and realizing its full potential when political events added an international dimension to the exploration process and provided new and highly profitable

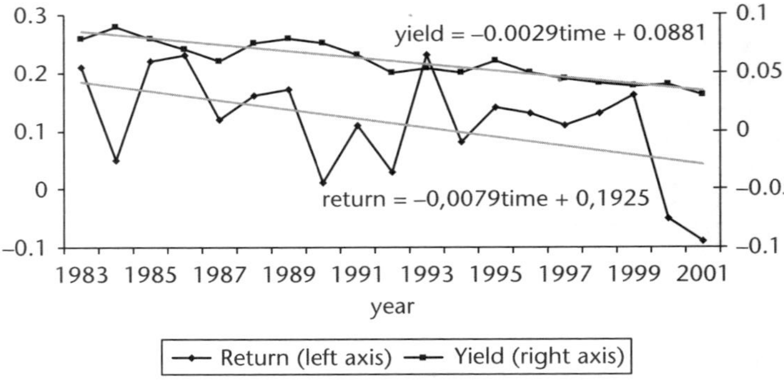

Figure 2.4 Returns on US foreign assets
Source: Lane and Milesi-Feretti (2003: 31, figure 10).

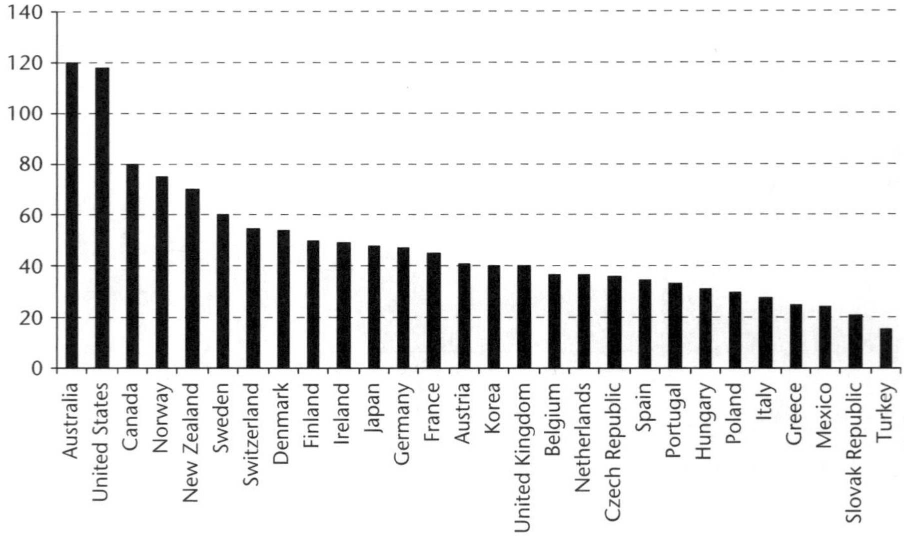

Figure 2.5 PCs per 100 white-collar workers
Source: OECD (2005).

opportunities for application. It is well established that ICT can be considered a GPT (in fact the concept was developed with ICT in mind by Bresnahan and Trajtenberg 1995). As an illustration Figure 2.5 shows that by 2000 most OECD countries had over 40 per cent of their workforce using computers.

This establishes

Fact 3: *ICT technology is a pervasive GPT.*

We conclude that the drop in political risk, the expansion of the low-skilled labour force in the South and the introduction of a new GPT are the fundamental exogenous events that triggered structural change in the last quarter of the twentieth century. They all created new opportunities for Northern entrepreneurs who took a leading role in shaping global production and trade. Their response to these new opportunities is what effectuated the changes we set out to explain. Before proceeding with modelling their role, however, we first establish those changes as empirically relevant facts in North and South.

2.2.2 *Shifting international specialization and the industrial economy in the South (East)*

A first indication of changing patterns of global economic activity is given by the more traditional measures of transnational activity. These traditional measures include trade (exports and imports), FDI (inward), international capital flows, and inter-country labour mobility. The overall trend for all of these measures has been strongly positive.

Both exports and imports have become more important and accounted for a greater share of economic activity. Total world exports increased from US$1.3 trillion in 1970 to nearly US$5 trillion in 2000, in constant dollars. Exports have increased their share in total output consistently since 1820 (Torres 2001: table 1), but as Figure 2.6 shows the increase in shares has also accelerated since the 1960s.[2] Exports matter absolutely, relatively, and increasingly.

Of course higher total exports could simply be more of the same products by the same players crossing more borders. But this is not the case. The economic structure has fundamentally changed over the past 25 years in the developing South. Where the South previously relied on local production, agricultural,

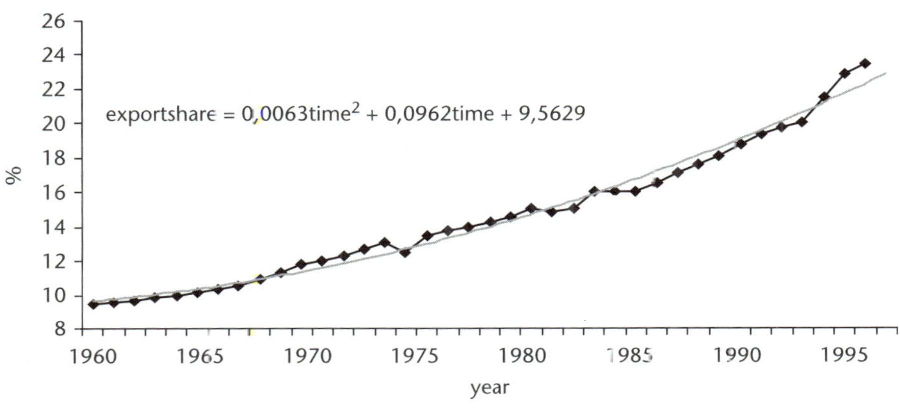

Figure 2.6 Share of exports in world GDP
Source: Torres (2001: 11, figure 1).

and natural resources exports while importing most manufactures, they now increasingly compete for a share in the global manufactures and even services markets. Especially in South East Asia, where land/labour ratios are low, manufacturing, not agriculture and/or mining, is the key sector of economic growth. Mayer and Wood (1999) predicted that the large economies of China and India would grow towards manufacturing shares of between 70 and 80 per cent of exports and by the mid-nineties these countries had already reached those levels. China and India are by far the most impressive examples of rapid industrialization in the region but as OECD (2005) shows the overall non-OECD share in world trade and manufactures is increasing steadily.

A closer look at the data shows that it is especially intermediate industrial products from South East Asia that are responsible for this upward trend (see Feenstra and Hanson 2004; WTO 2005). In several countries in South East Asia and some countries in South America, industrial employment and output shares rose as a result of this outsourcing wave and by the turn of the last century various industrialization processes have gained momentum to the point that it is fair to speak of emerging industrial societies. In Taiwan and South Korea for example, employment in agriculture has dropped to around 10 per cent, a level comparable to that of New Zealand today and half that of the Netherlands in 1950. The large populous economies in the East are also rapidly industrializing and with over 50 per cent of the workforce still in agriculture, they have a huge industrial labour reserve yet to be absorbed.

Evidence presented by Antras (2005) suggests that Northern investors and entrepreneurs played a vital role in fuelling these waves of industrialization.[3] International flows of capital, and in particular of FDI towards these new kids on the industrialized block, have exploded since the late 1980s and have

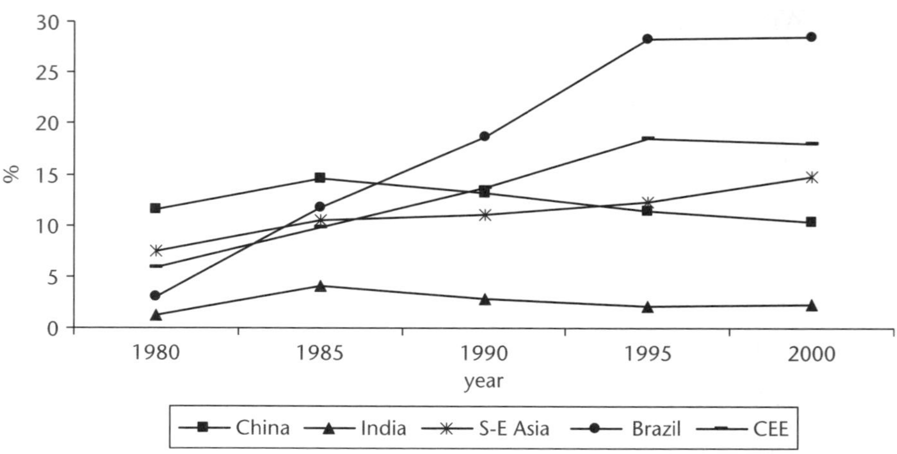

Figure 2.7 FDI inflow as percentage of gross fixed capital formation
Source: UNCTAD (2003).

increasingly substituted for domestic savings as a source of gross fixed capital formation (see Figure 2.7).

In general it is fair to say that the rise of industrial production in the South, at least in its initial stages, was actively promoted by actions of economic agents from the North. We thus establish

Fact 4: *Northern entrepreneurs fuelled South (East)ern industrialization.*

2.2.3 *Shifting labour demand and the entrepreneurial economy in the North*

In the North, meanwhile, traditional industrial employment in large industrial firms has been falling as production was relocated. However, overall employment and output have still grown, mainly due to small firm job creation in ICT-related services. The well-known transition towards the services economy is clearly illustrated in Figure 2.8.

In particular the widening gap between industry and services since 1970 indicates an unprecedented labour reallocation. Arguably, for the first time, reallocation proceeded in a time span that is considerably shorter than the average career, making life-long employment a rarity and firm- and industry-specific skills a more risky investment than before.

A closely related phenomenon is the rapid increase in the knowledge-intensity of production, both in industry and services in the North. The trends away from low-skilled manufacturing in OECD labour markets have been well documented indeed (Juhn, Murphy, and Pierce 1993; Gottschalk and Moffit 1994; Bell 1996; Katz and Autor 1999; Goldin and Katz 2001;

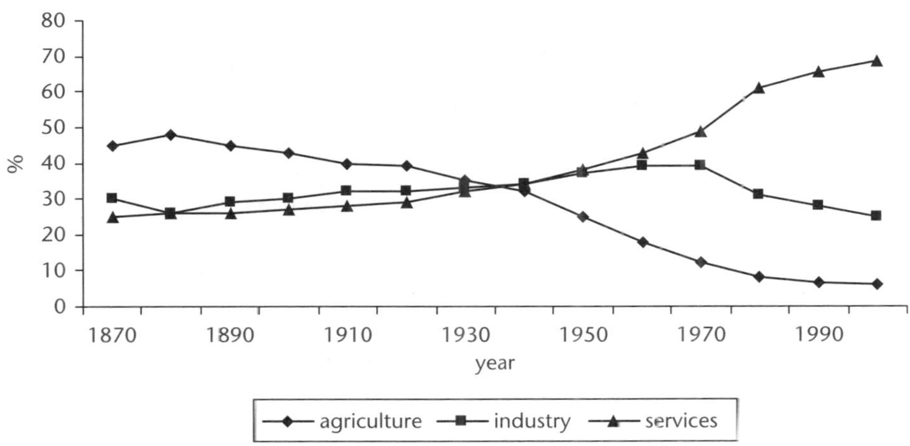

Figure 2.8 Employment shares in 17 OECD countries

Source: Iversen and Rosenbluth (2006, figure 4).

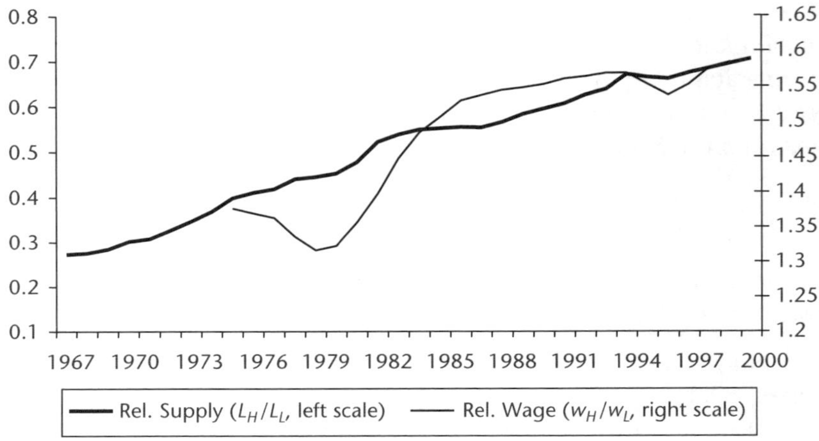

Figure 2.9 Relative wage and supply US 1967–2000

Source: CPS March data and computations at <http://www.bls.census.gov/cps/ads/adsmain.htm>, following Card and DiNardo (2002).

Nickell and Acemoglu 2002a). Figure 2.9 shows how relative wages have been decreasing in the US for low-skilled workers in spite of strong increases in the average educational level, suggesting massive relative demand shifts away from low-skilled labour.

A similar demand trend, even if wage changes were often less dramatic, has been identified for virtually every OECD country (see for example Autor, Katz, and Kearney (2005) for recent evidence on the US and Machin and Van Reenen (1998) or Sanders and Ter Weel (2000) for surveys of the evidence on the OECD).

It has been argued that technology itself or the increased rate of innovation has produced this skill bias in labour demand (e.g. Sanders 2005). Greenwood and Uysal (2004) make the case for the acceleration hypothesis compellingly. The number of firms per capita and trademarks granted has exploded over the 1980s in the US. Kortum and Lerner (1997) show a similar speed-up in patenting and most activity obviously takes place in the new technology fields such as biotech and software. All this increased activity indicates high levels of grassroots entrepreneurship and product innovation, which in Greenwood (1997) but also in life cycle models causes temporary shifts in the relative demand for skilled labour.

Meanwhile changing trade patterns also affected the North. The trade of goods tripled between 1985 and 1996, while the trade of services increased by more than three times over this time period. While some of this increase in the world export rate is attributable to an increased participation in international trade by countries, which had previously been excluded, export rates in the leading industrialized countries have also increased over the past three decades. For example, US exports and imports have increased from 11 per cent of GDP in 1970 to more than 25 per cent by 1999.[4] Exports as a share of GDP for 49 other developed countries have risen from around 18 per cent in 1982 to around 25 per cent by 1999.

Evidence, collected by Feenstra and Hanson (1996, 1999, 2004), shows that over the 1980s and 1990s the import of intermediate (manufactured) goods has almost doubled from 6.5 to 11.6 per cent of total intermediate purchases in the US. More open economies are likely to show much higher levels and similar increases.

The way we read the evidence is that globalization combined with technological change, and in particular the information technology and communication technologies breakthroughs, had rendered the comparative advantage in low-technology and even traditional moderate technology industries incompatible with high wage-levels. At the same time, the emerging comparative advantage that is compatible with high wage-levels is based on innovative and knowledge-intensive activities. There are many indicators reflecting the shift in the comparative advantage of the high-wage countries towards an increased importance of such activities. Before we turn to the role of the entrepreneur in this 'new' economy we establish

> Fact 5: *The North has experienced deindustrialization and a changing relative labour demand towards skilled white-collar workers.*

Increased globalization of economic activity also seemed to condemn entrepreneurship, in the form of new firm start-ups and small firms, to extinction. As Chandler (1990: 130) concluded, 'to compete globally you have to be big'. Thus, it was particularly startling and a seeming paradox, when scholars first began to document that what had looked like the inevitable demise of entrepreneurship actually began to reverse itself from the 1970s onwards.

Loveman and Sengenberger (1991) and Acs and Audretsch (1993) documented the reemergence of small firms and entrepreneurship in North America and Europe. In 1981 David Birch revealed the startling findings from his long-term study of US job generation. Birch (1981: 8) found that: 'whatever else they are doing, large firms are no longer the major providers of new jobs for Americans'. Instead, he discovered that most new jobs emanated from small firms. Davis, Haltiwanger, and Schuh (1996a, 1996b) recently confirmed Birch's results and indicate that small firms are net job generators, while large firms shed labour. The rise in entrepreneurial activity is not particular to the US. An EIM study documents how the relative importance of SMEs in Europe (19 countries), in terms of employment shares, has continued to increase between 1988 and 2001 (EIM 2002).

Hence, the weight of the empirical evidence on employment generation is remarkably robust and indicates that the role of entrepreneurship in employment generation in Europe is consistent with the findings for the US. The reversal of the trend towards large enterprises and the reemergence of small firms were not limited to employment only. For example, in The Netherlands the business ownership rate fell during the post-war period, until it reached a trough of 0.085 in 1982. But this downward trend was subsequently reversed, rising to a business ownership rate of 0.10 by 1998 (Audretsch et al. 2002).

Small firms have also increased their share in value added. In the US the average real GDP per firm increased by nearly two-thirds between 1947 and 1980, from US$150,000 to US$245,000, reflecting a trend towards larger enterprises but within the subsequent seven years, by 1987, it had fallen by about 14 per cent to US$210,000 (Brock and Evans 1989). Similarly, small firms accounted for one-fifth of manufacturing sales in the US in 1976, but by 1986 this share had risen to over one-quarter (Acs and Audretsch 1990).

Small and new enterprises are therefore increasingly important qualitatively, as a mode of organization and operation, and quantitatively as an engine of employment creation on both sides of the Atlantic. At this point we feel confident to state

Fact 6: *Entrepreneurial activity has risen in the North.*

2.2.4 *Summary*

This section has established the facts we intend to explain and those we intend to use as explanation. We intend to explain how political liberalization and technology have interacted and offered the Northern entrepreneurs the opportunities they needed to transform trade and production patterns and shape what we know as globalization and the knowledge economy. We have shown that:

1. Labour abundant countries have joined the global economic system.
2. Political risk has declined in the world.
3. ICT is a GPT that fuels globalization.

These developments are exogenous to the economic system. We posit that they may cause:

4. FDI-driven industrialization in the South (East).
5. Deindustrialization and shifting relative labour demand in the North.
6. Increased entrepreneurial activity in innovation, outsourcing, and offshoring.

2.3 The model

Aspects involving the decision to locate production in foreign countries are certainly not a new subject and have been studied in the context of so-called product life cycle models since Vernon coined the term in 1966. He linked the location of production activities to the life cycle stage of the product, hypothesizing that new products, because of their higher knowledge-intensity, would be produced in the North, which has a relative abundance of skilled labour and therefore a comparative advantage in their production. Krugman (1979) formalized Vernon's idea in a model but assumed that technology generation and transfer are both exogenous processes. Grossman and Helpman (1991) used the insights of endogenous growth theories to endogenize the process of innovation and imitation. In these models, however, no distinction is made between skilled and unskilled labour. Moreover, Grossman and Helpman are also not explicit about the role of Northern agents in initiating the transfer of technology across geographic space. Instead they assume that R&D undertaken in the South drives technology diffusion by copying ideas off the shelf from the North.

In this tradition we now present a model that distinguishes among three stages in the life cycle of products. The first stage is initiated by the discovery of a new product. Upon discovery, an entrepreneur in the North will introduce the new product or service into the Northern economy. The second stage occurs when the product matures, leading the incumbent firm to apply firm-based R&D to the standardization and routinization of the production process, driven by the desire to cut costs. As the production process becomes standardized, less skilled workers in the North can then be engaged in the production process at relatively low wages. However, so can Southern workers. Still, we assume that the act of moving production abroad marks the beginning of a separate third and final stage in the product life cycle. Setting up production facilities in the South is neither costless nor riskless and requires an entrepreneurial act. We assume that a Northern entrepreneur initiates this process as copying behaviour is important especially after Northern entrepreneurs have set up shops in the South. We therefore assume that it is FDI, not Southern imitation R&D that will be the medium of knowledge spillover.

By endogenizing the allocation of entrepreneurial talent over product innovation, process innovation, and outsourcing/off-shoring, we can explain the dynamics in global specialization patterns as the result of political risks falling and low-skilled labour abundant countries joining the global economy. We also show that the introduction of a so-called GPT drives the Northern economy further towards what we have called the entrepreneurial economy above; a production structure that relies heavily on new goods production and product innovation for creating value added.

The model structure follows that of a standard endogenous growth with variety expansion in a final goods model. First we assume identical time separable preferences globally and derive consumer demand for goods over time solving:

$$\max_{E_t} : \int_t^{\infty} e^{-\rho(\tau-t)} \log(E_\tau/P_\tau)d\tau \qquad (1)$$
$$s.t. : Y_t + rA_t = E_t + \dot{A}_t$$

where E is expenditure on consumption, P is the price of a unit of direct utility, ρ is the discount rate, Y is income, A is the level of assets and a dot over a variable signifies the time derivative. This problem yields the familiar Ramsey result:

$$\dot{E}_t/E_t = r - \rho \qquad (2)$$

Then we assume a standard Constant Elasticity of Substitution (CES) love of variety instant utility function and solve:

$$\max_{c_i} : U = \left(\int_0^n c_i^\alpha di\right)^{1/\alpha} \quad \text{where } 0 \leq \alpha \leq 1 \qquad (3)$$
$$s.t. : \int_0^n c_i p_i di \leq E$$

i indexes a single good and n is the total number of goods consumed. c and p are quantity and price of a single good in the basket. To derive the instant global demand functions for all current and future goods in this CES utility function is straightforward:[5]

$$c^D{}_i = \left(\frac{p_i}{P}\right)^{\frac{1}{\alpha-1}} \frac{E}{P} \text{ where } p \equiv \left(\int_0^n p_{-i}^{\frac{\alpha}{\alpha-1}} di\right)^{\frac{\alpha-1}{\alpha}} \qquad (4)$$

P is defined as the minimum cost of one unit of direct utility U.

Now consider production. Following the life cycle literature we assume that products in their early stages can only be produced in the North. The reason is that early stage production requires high flexibility and creativity as well as frequent feedback from the consumer. Northern skilled workers and sophisticated consumers provide the required setting. All these early stage aspects give skilled workers a competitive advantage in production (Schultz 1961). As there is also a large tacit knowledge spillover required from inventor to innovator and early stage producer, it is reasonable to assume that a new product is initially produced only by skilled workers and only in the North.

At that stage the producer therefore faces no competition, at home or abroad, allowing him to set monopoly prices:

$$\begin{aligned} \max_{p_i} &: \pi_i = c_i p_i - w^N{}_H l^N{}_{Hi} \\ s.t. &: c_i = c^D{}_i \\ s.t. &: y_i = b^N{}_H l^N{}_{Hi} \end{aligned} \quad (5)$$

where $w^N{}_H$ is the high-skilled wage in the North, l is the amount of labour employed. Note also that labour is the only factor of production and the production volume y is assumed linear in labour employed. This yields:

$$p_i = \frac{w^N{}_H}{\alpha b^N{}_H} \; \forall n \in n^N{}_H \quad (6)$$

As a product matures the incumbent producer will invest in R&D to come up with a process that allows the product to be produced by low-skilled workers in the North.[6] This will reduce costs and hence increases profits. At the mature stage, however, the market will no longer be uncontested. The fired skilled workers can always set up a new firm and set their price equal to marginal costs:[7]

$$p_i = \frac{w^N{}_H}{b^N{}_L} \quad (7)$$

The incumbent, however, using the mature production technology, has lower marginal costs and can retain the entire market and prevent entry by setting his price equal to:

$$p_i = \min\left[\frac{w^N{}_L}{\alpha b^N{}_L}, \frac{w^N{}_H}{b^N{}_H} - \epsilon\right] \quad (8)$$

where ε is close to 0. However, the new production process also creates the possibility of outsourcing the production. To enable Northern low-skilled workers to compete against the high-skilled workers, production must be routinized and many routines are embedded into machines, interfaces, and procedures. Such codifications, however, can also be transferred (at some cost) abroad. The Northern producer must therefore now fear competition from abroad. His only edge is the potentially higher marginal productivity of Northern low-skilled labour.[8] If profitable entry from abroad is feasible the incumbent is not likely to engage in that activity. It typically takes an entrepreneur to set it up, take the risks and, if successful, reap the benefits. Products that have been outsourced have a price:

$$p_i = \min\left[\frac{w^S{}_L}{\alpha b^S{}_L}, \frac{w^N{}_L}{b^N{}_L} - \epsilon\right] \quad (9)$$

and in our model compete out production in the North entirely. For given n, $n^N{}_H$, $n^N{}_L$, and $n^S{}_L$ the labour markets for Northern skilled, unskilled, and Southern labour can now be assumed to clear, such that the model equilibrates. We require:

$$
\begin{aligned}
L^N{}_H* &= \int_{n^N{}_L+n^S{}_L}^{n} \alpha^{\frac{1}{1-\alpha}} b^{\frac{\alpha}{1-\alpha}} EP^{\frac{\alpha}{1-\alpha}} (w^N{}_H)^{\frac{1}{\alpha-1}} di = n^N{}_H \left(\alpha^{\frac{1}{1-\alpha}} b^{\frac{\alpha}{1-\alpha}} EP^{\frac{\alpha}{1-\alpha}} (w^N{}_H)^{\frac{1}{\alpha-1}} \right) \\
L^N{}_L* &= \int_{n^S{}_L}^{n-n^H{}_L} b^{-1} EP^{\frac{\alpha}{1-\alpha}} \left(\min\left[\frac{w^N{}_H}{b}, \frac{w^N{}_L}{\alpha b} \right] \right)^{\frac{1}{\alpha-1}} di = n^N{}_L \left(b^{-1} EP^{\frac{\alpha}{1-\alpha}} \left(\min\left[\frac{w^N{}_H}{b}, \frac{w^N{}_L}{\alpha b} \right] \right)^{\frac{1}{\alpha-1}} \right) \\
L^S{}_L* &= \int_{0}^{n^S{}_L} b^{-1} EP^{\frac{\alpha}{1-\alpha}} \left(\min\left[\frac{w^N{}_L}{b}, \frac{w^S{}_L}{\alpha b} \right] \right)^{\frac{1}{\alpha-1}} di = n^S{}_L \left(b^{-1} EP^{\frac{\alpha}{1-\alpha}} \left(\min\left[\frac{w^N{}_L}{b}, \frac{w^S{}_L}{\alpha b} \right] \right)^{\frac{1}{\alpha-1}} \right)
\end{aligned}
\tag{10}
$$

Although this set of equations cannot be solved analytically, due to the discontinuities in the demand curves, one can analyse the equilibrium graphically and sequentially. First, consider the market for high-skilled workers in the North in the left panel.

There is no competition in this market, implying that prices are always set to profit maximization levels and consequently the demand curve is continuous. Confronted with exogenous supply this yields an equilibrium wage at $w^N{}_H*$. Given that wage the marginal costs in the new goods sector are given by $w^N{}_H*/b$, which puts an upper limit on prices in the mature Northern sector. Beyond that price the demand for goods falls to 0, so demand for labour falls to 0. Therefore the demand curve has a horizontal section at the low-skilled wage

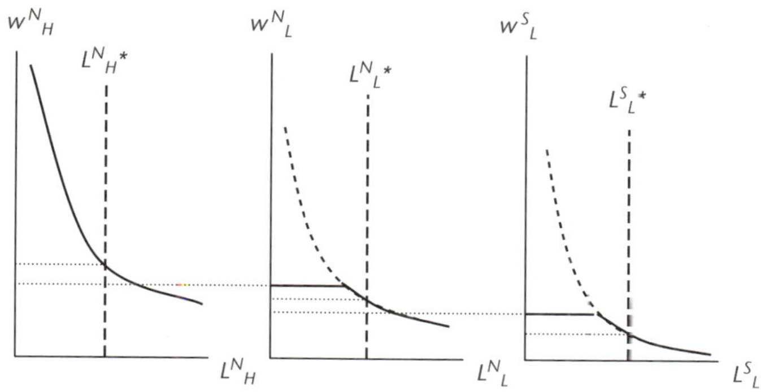

Figure 2.10 Labour market equilibrium

Source: compiled by the authors.

that would imply a profit maximizing price is set equal to $w^N{}_H{}^*/b$. This is the case at $w^N{}_L = \alpha w^N{}_H{}^*$. Then the kinked curve is confronted with exogenous supply and we obtain the Northern equilibrium low-skilled wage. By similar argument one can also construct the demand for Southern low-skilled labour and solve the final market. All labour markets equilibrate simultaneously for all given combinations of n, $n^N{}_H$, $n^N{}_L$, and $n^S{}_L$. n, $n^N{}_H$, $n^N{}_L$, and $n^S{}_L$, however, will not be constant over time. The assumption of imperfect competition implies there are rents in the model and therefore there is an incentive for new entrants.

Consider the effects on profit flows and ultimately the value of being a producer in the three stages of the life cycle. With prices set as above, there are four scenarios that we need to consider. The cost gaps between high- and low-skilled labour using producers in the North and between Northern and Southern low-skilled labour using producers can be wide or narrow respectively. Hence profits for the three types of activity under these conditions are given in Table 2.1. The most important result in Table 2.1 is the fact that profits are always positive and in steady-state equilibrium, where relative wages and relative sector sizes are constant, proportional to total expenditures, and inversely proportional to their own sector size. Given these profit flows it is therefore worthwhile to engage in entrepreneurial activity, product development, and outsourcing/off-shoring.[9]

Entrepreneurial activity introduces new goods and services into the economy. Product development allows existing products to be produced by low-skilled labour in the North and outsourcing moves the production of routinized products to the South. Even if patents are perfectly enforced, such activities are worthwhile, as the flow of profit increases with each stage in the life cycle, as long as $w^N{}_H > w^N{}_L > w^S{}_L$. The incentive to move the product over the cycle then comes from increments in the profit flow. This also explains why incumbents may engage in (defensive) product development and outsourcing even though they cannibalize on their current profits.[10] The

Table 2.1 Profits under various scenarios

$n^N{}_L$-	Gap $n^N{}_H$-$n^N{}_L$	
$n^S{}_L$	Wide	Narrow
Wide	$\pi^N{}_H = \frac{(1-\alpha)E}{n^N{}_H}\left(1+\frac{n^N{}_L}{n^N{}_H}\left(\frac{w^N{}_L}{w^N{}_H}\right)^{\frac{\alpha}{\alpha-1}}\left(\frac{b^N{}_L}{b^N{}_H}\right)^{\frac{\alpha}{1-\alpha}}+\frac{n^S{}_L}{n^N{}_H}\left(\frac{w^S{}_L}{w^N{}_H}\right)^{\frac{\alpha}{\alpha-1}}\left(\frac{b^S{}_L}{b^N{}_H}\right)^{\frac{\alpha}{1-\alpha}}\right)^{-1}$ $\pi^N{}_L = \frac{(1-\alpha)E}{n^N{}_L}\left(1+\frac{n^N{}_H}{n^N{}_L}\left(\frac{w^N{}_H}{w^N{}_L}\right)^{\frac{\alpha}{\alpha-1}}\left(\frac{b^N{}_H}{b^N{}_L}\right)^{\frac{\alpha}{1-\alpha}}+\frac{n^S{}_L}{n^N{}_L}\left(\frac{w^S{}_L}{w^N{}_L}\right)^{\frac{\alpha}{\alpha-1}}\left(\frac{b^S{}_L}{b^N{}_L}\right)^{\frac{\alpha}{1-\alpha}}\right)^{-1}$ $\pi^S{}_L = \frac{(1-\alpha)E}{n^S{}_L}\left(1+\frac{n^N{}_L}{n^S{}_L}\left(\frac{w^N{}_L}{w^S{}_L}\right)^{\frac{\alpha}{\alpha-1}}\left(\frac{b^N{}_L}{b^S{}_L}\right)^{\frac{\alpha}{1-\alpha}}+\frac{n^N{}_H}{n^S{}_L}\left(\frac{w^N{}_H}{w^S{}_L}\right)^{\frac{\alpha}{\alpha-1}}\left(\frac{b^N{}_H}{b^S{}_L}\right)^{\frac{\alpha}{1-\alpha}}\right)^{-1}$	$\pi^N{}_H = \frac{(1-\alpha)E}{n^N{}_H}\left(1+\frac{n^N{}_L}{n^N{}_H}\alpha^{\frac{\alpha}{\alpha-1}}+\frac{n^S{}_L}{n^N{}_H}\left(\frac{w^S{}_L}{w^N{}_H}\right)^{\frac{\alpha}{\alpha-1}}\left(\frac{b^S{}_L}{b^N{}_H}\right)^{\frac{\alpha}{1-\alpha}}\right)^{-1}$ $\pi^N{}_L = \frac{(1-\alpha)E}{n^N{}_L}\left(1+\frac{n^N{}_H}{n^N{}_L}\alpha^{\frac{\alpha}{1-\alpha}}+\frac{n^S{}_L}{n^N{}_L}\left(\frac{w^S{}_L}{w^N{}_H}\right)^{\frac{\alpha}{\alpha-1}}\left(\frac{\alpha b^S{}_L}{b^N{}_H}\right)^{\frac{\alpha}{1-\alpha}}\right)^{-1}$ $\pi^S{}_L = \frac{(1-\alpha)E}{n^S{}_L}\left(1+\frac{n^N{}_L}{n^S{}_L}\left(\frac{w^N{}_H}{w^S{}_L}\right)^{\frac{\alpha}{\alpha-1}}\left(\frac{b^N{}_H}{\alpha b^S{}_L}\right)^{\frac{\alpha}{1-\alpha}}+\frac{n^N{}_H}{n^S{}_L}\left(\frac{w^N{}_H}{w^S{}_L}\right)^{\frac{\alpha}{\alpha-1}}\left(\frac{b^N{}_H}{b^S{}_L}\right)^{\frac{\alpha}{1-\alpha}}\right)^{-1}$
Narrow	$\pi^N{}_H = \frac{(1-\alpha)E}{n^N{}_H}\left(1+\frac{n^N{}_L}{n^N{}_H}\left(\frac{w^N{}_L}{w^N{}_H}\right)^{\frac{\alpha}{\alpha-1}}\left(\frac{b^N{}_L}{b^N{}_H}\right)^{\frac{\alpha}{1-\alpha}}+\frac{n^S{}_L}{n^N{}_H}\left(\frac{w^N{}_L}{w^N{}_H}\right)^{\frac{\alpha}{\alpha-1}}\left(\frac{b^S{}_L}{\alpha b^N{}_H}\right)^{\frac{\alpha}{1-\alpha}}\right)^{-1}$ $\pi^N{}_L = \frac{(1-\alpha)E}{n^N{}_L}\left(1+\frac{n^N{}_H}{n^N{}_L}\left(\frac{w^N{}_H}{w^N{}_L}\right)^{\frac{\alpha}{\alpha-1}}\left(\frac{b^N{}_H}{b^N{}_L}\right)^{\frac{\alpha}{1-\alpha}}+\frac{n^S{}_L}{n^N{}_L}\alpha^{\frac{\alpha}{1-\alpha}}\right)^{-1}$ $\pi^S{}_L = \frac{(1-\alpha)E}{n^S{}_L}\left(1+\frac{n^N{}_L}{n^S{}_L}\alpha^{\frac{\alpha}{1-\alpha}}+\frac{n^N{}_H}{n^S{}_L}\left(\frac{w^N{}_H}{w^N{}_L}\right)^{\frac{\alpha}{\alpha-1}}\left(\frac{\alpha b^N{}_H}{b^N{}_L}\right)^{\frac{\alpha}{1-\alpha}}\right)^{-1}$	$\pi^N{}_H = \frac{(1-\alpha)E}{n^N{}_H}\left(1+\frac{n^N{}_L}{n^N{}_H}\alpha^{\frac{\alpha}{\alpha-1}}+\frac{n^S{}_L}{n^N{}_H}\left(\frac{w^N{}_L}{w^N{}_H}\right)^{\frac{\alpha}{\alpha-1}}\left(\frac{b^N{}_L}{\alpha b^N{}_H}\right)^{\frac{\alpha}{1-\alpha}}\right)^{-1}$ $\pi^N{}_L = \frac{(1-\alpha)E}{n^N{}_L}\left(1+\frac{n^N{}_H}{n^N{}_L}\alpha^{\frac{\alpha}{1-\alpha}}+\frac{n^S{}_L}{n^N{}_L}\left(\frac{w^N{}_L}{w^N{}_H}\right)^{\frac{\alpha}{\alpha-1}}\left(\frac{b^N{}_L}{b^N{}_H}\right)^{\frac{\alpha}{1-\alpha}}\right)^{-1}$ $\pi^S{}_L = \frac{(1-\alpha)E}{n^S{}_L}\left(1+\frac{n^N{}_L}{n^S{}_L}\left(\frac{w^N{}_H}{w^N{}_L}\right)^{\frac{\alpha}{\alpha-1}}\left(\frac{b^N{}_H}{b^N{}_L}\right)^{\frac{\alpha}{1-\alpha}}+\frac{n^N{}_H}{n^S{}_L}\left(\frac{w^N{}_H}{w^N{}_L}\right)^{\frac{\alpha}{\alpha-1}}\left(\frac{\alpha b^N{}_H}{b^N{}_L}\right)^{\frac{\alpha}{1-\alpha}}\right)^{-1}$

incentive to become an innovating entrepreneur is the value of being a Northern uncontested new good producer that is given by:

$$v^N{}_H = E\left\langle \int_t^T e^{-r(\tau-t)} \pi^N{}_H(\tau) d\tau \middle| I(t) \right\rangle \tag{11}$$

the expected discounted profit flow from time t, the time of entry, to time T, the time at which production with low-skilled labour starts and monopoly profits are contested, conditional on the information set $I(t)$ available at time t. The flow probability of being contested is given by the relative gross number of product development, $(\dot{n}^N{}_L + \dot{n}^S{}_L)/n^N{}_H$. Expecting that profit grows at a constant rate, as it will in the steady-state, yields:

$$\begin{aligned} v^N{}_H &= \pi^N{}_H(t) \int_t^\infty e^{-(r+(\dot{n}^N{}_L+\dot{n}^S{}_L)/n^N{}_H)(\tau-t)} \pi^N{}_H(t) e^{(\dot{E}/E-\dot{n}^N{}_H/n^N{}_H)(\tau-t)} \\ d\tau &= \frac{\pi^N{}_H(t)}{r - \dot{E}/E + \dot{n}/n^N{}_H} \end{aligned} \tag{12}$$

An entrepreneur is able to capture this value by setting up a firm and starting production of a new good or service. Similarly we have for the Northern low-skilled producers:

$$v^N{}_L = \frac{\pi^N{}_L(t)}{r - \dot{E}/E + (\dot{n}^S{}_L + \dot{n}^N{}_L)/n^N{}_L} \tag{13}$$

where the current profits are discounted by the rate of interest r, the rate of profit erosion $-(\dot{E}/E - \dot{n}^N{}_L/n^N{}_L)$, and the risk of losing the profits because of outsourcing, $\dot{n}^S{}_L/n^N{}_L$. However, the value of a product development to the incumbent is lower. In order to obtain the higher profit flow an incumbent foregoes the existing profit flow. Hence the value of a product development equals the positive difference between (13) and (12). Finally, the value of owning an outsourced firm that produces in the South is equal to:

$$v^S{}_L = \frac{\pi^S{}_L(t)}{r + \psi - \dot{E}/E + \dot{n}^S{}_L/n^S{}_L} \tag{14}$$

where ψ is a risk premium for producing abroad that reflects the risks of being unable to retrieve profits and reflects the perhaps less than perfect property rights protection in the South.[11] Once more the value to the incumbent and therefore also to any entrant that is forced to compensate him for patent infringement equals the positive difference between (14) and (13).[12]

Let us finally assume that introducing a new good or service, developing a product, and outsourcing it all draw on the same pool of limited resources in the North, which we may label entrepreneurial talent T.[13] To close the model we must specify how entrepreneurial talent combines with accumulated knowledge to generate the dynamics in the model. Assume:

$$\begin{aligned} \dot{n}^{N}{}_{H} &= \dot{n} - \dot{n}^{N}{}_{L} - \dot{n}^{S}{}_{L} = aT_I^{\gamma}\, n^{A} - \dot{n}^{N}{}_{L} - \dot{n}^{S}{}_{L} \\ \dot{n}^{N}{}_{L} &= aT_D^{\gamma}\, n^{N}{}_{H} - \dot{n}^{S}{}_{L} \qquad \text{where } 0 \leq \gamma \leq 1 \qquad (15) \\ \dot{n}^{S}{}_{L} &= aT_O^{\gamma}\, {}_{n}{}^{LN} \end{aligned}$$

Note that a productivity parameter a and the amount of entrepreneurial talent allocated are combined with n^A, $n^N{}_H$, and $n^N{}_L$ to reflect the knowledge spillovers in the model. Innovation benefits from the (exogenous) stock of knowledge n^A.[14] Development benefits from the knowledge accumulated in new goods production $n^N{}_H$ and outsourcing benefits from experience in routinized production $n^N{}_L$. Furthermore (15) reflects the assumption that a product must first be invented, can then be routinized, and finally outsourced, but only in that order, missing none of the stages in its development.

Entrepreneurial talent now has to be allocated over three alternative uses. In each it must generate the same marginal value product. Assuming this (market clearing) reward to entrepreneurial talent is given by w_T we obtain:

$$w_T = \gamma a T_I^{\gamma-1} n^A v^N{}_H = \gamma a T_D^{\gamma-1} n^N{}_H (v^N{}_L - v^N{}_H) = \gamma a T_O^{\gamma-1} n^N{}_L (v^S{}_L - v^N{}_L) \qquad (16)$$

This arbitrage condition can be rewritten in separate entrepreneurial talent demand equations that are downward sloping in the reward for entrepreneurial talent.[15] Equating the sum to the total stock of entrepreneurial talent in the North yields the equilibrium reward for entrepreneurial activity:

$$w_T = aT^{*\gamma-1} n^A v^N{}_H \left(1 + \left(\frac{n^N{}_H}{n^A} \left(\frac{v^N{}_L}{v^N{}_H} - 1 \right) \right)^{\frac{1}{1-\gamma}} + \left(\frac{n^N{}_L}{n^A} \left(\frac{v^S{}_L}{v^N{}_H} - \frac{v^N{}_L}{v^N{}_H} \right) \right)^{\frac{1}{1-\gamma}} \right)^{\gamma(1-\gamma)} \qquad (17)$$

which in steady-state grows in proportion to n^A times $v^N{}_H$. As the latter grows in proportion to profits, which in turn grow in proportion to expenditure E over $n^N{}_H$ in the steady-state the reward to entrepreneurial talent will grow in proportion to total expenditure, as will wages in all labour markets, as long as n^A grows at the same rate as $n^N{}_H$. Normalization of expenditure then pins down the growth rate of all nominal variables in the model to 0. The total range of goods expands at the common growth rate of new, mature, and outsourced goods in the global economy. This leads to propositions 1 and 2.

Proposition 1:

There exists a steady-state equilibrium in which the allocation of all labour types is stable, all ranges of goods grow at the common positive rate, and labour markets are clear.

Proposition 2:
This steady-state is a unique and globally stable equilibrium in the sense that the economy will converge back to this steady-state when it is disturbed by shocks.

The proof for both propositions is in the Appendix. The dynamic equilibrium in the model can be illustrated by plotting the three demand equations for entrepreneurial activity, implicit in equation (16) and equating the vertical sum of these curves to the exogenous supply of entrepreneurial talent in the North. In the steady-state these levels of activity must generate the same rate of expansion G to the respective variety ranges. The knowledge spillover structure assumed in equation (15), which implies all entrepreneurial activity receives a spillover from upstream entrepreneurial activity (which for innovation is exogenous knowledge creation), guarantees that in the end all variety ranges will grow at the same rate for a stable allocation of entrepreneurial talent. As was shown under Proposition 2 and in the Appendix this implies that in the steady-state the right panel of Figure 2.11 applies.

Now consider the comparative statics in the steady-state concerning the effects of increases in $L^S{}_L$ (globalization/integration), ψ (property rights/institutions), n^A (technology shock/GPT), and our policy experiment to encourage and stimulate entrepreneurship T^* (education/entrepreneurship). To see what happens consider first the shocks in isolation.

A sufficiently large increase in Southern labour supply (if the shock is too small one may stay in a situation of narrow wage gap and relative profits do not change, causing no response) will initially drive down equilibrium wages in the South and push the Southern economy out of a possible narrow gap situation. In Table 2.1 it can be seen that in that case profits in the South will rise as wages fall. By equation (14) that implies the value of outsourcing activities increases and entrepreneurs will switch out of innovation and development to outsource more mature products to the South. As the rate of outsourcing accelerates and the rates of innovation and development fall, there is a change in the relative composition of the economy. More varieties are produced in the South, variety expansion falls and both $n^N{}_H$ and $n^N{}_L$ will fall relative to $n^S{}_L$. This, by (10) implies that Southern wages recover and Northern wages will fall, whereas the reward to entrepreneurial talent rises. Of course equilibrium is reestablished once diminishing returns in innovation, development, and outsourcing (through the gradual reduction of upstream knowledge stocks) make sure all ranges grow at the same rate again. It should be noted here that the prediction that Southern wages fall and recover gradually does not apply for the countries that join world markets. They enter the market with wages below $w^S{}_L$ and see an immediate rise followed by a further increase in the wage-level. The predictions for Southern wage-levels therefore rather apply to, for example, countries like Brazil, Mexico, and Egypt than India or China.

The impact of a drop in political risk ψ enters our model directly in the return to outsourcing equation (14). Decreasing risk implies lower discount rates of given profits, making outsourcing more attractive to the entrepreneurs. The impact on the allocation of entrepreneurial talent and the relative composition of the economy is similar as above. However, there is no steady-state reduction in Southern wages. Instead the wages rise due to the increased demand. In the North the implications for relative wages and income are similar as before. The entrepreneurs gain and the low-skilled workers lose out most but high-skilled wages only fall relative to Southern wages.

A rise in the knowledge stock n^A or the rate of knowledge expansion has a temporary or permanent effect on the steady-state composition of the economy, respectively. If there is a level shock the benefits will dissipate gradually. The high-skilled workers in the North benefit first. Entrepreneurs move into innovation and abandon development and outsourcing. As the available stock of new products expands, however, the entrepreneurs will return to development and ultimately outsourcing. The steady-state has no changes, so the economy will eventually return to its initial equilibrium. If the growth rate has increased permanently that implies that in the steady-state all ranges of goods will grow at a higher rate, but also that the steady-state size of the new product range will be larger. Consequently the high-skilled in the North gain relative to the low-skilled in North and South and again the entrepreneurs benefit.

A rise in T^*, the stock of entrepreneurial talent/resources in the Northern economy, has a peculiar steady-state impact in this model. The entrepreneurs will be allocated to expanding the three variety ranges and these and all ranges

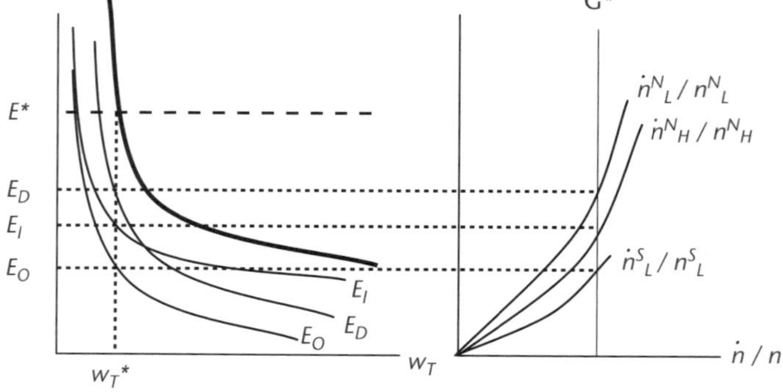

Figure 2.11 Steady-state equilibrium

Source: compiled by the authors.

will therefore expand faster for some time, but as the new and mature goods ranges are also eroded faster only the range of Southern varieties will grow in the steady-state. This generates the interesting result that the benefits of entrepreneurship training in general leak to developing countries. As entrepreneurs pass the buck, the buck moves faster to where it stops. However, one should realize that in passing the buck the entrepreneur appropriates part of the surplus with every pass. Entrepreneurial output thus increases and even though their wages fall, their incomes rise to the extent that Northern income increases relative to Southern income if demand elasticities are sufficiently low. One should also realize that the North is composed of many countries and any one Northern entrepreneur may take a mature good and outsource it. Having more entrepreneurs and fewer workers in the country gives a favourable position relative to the other Northern countries, not least because more entrepreneurs also means quicker adjustment to external shocks as the ones described above. From a global perspective more entrepreneurs are also unambiguously beneficial. It will reduce global wage if not income inequality and ensures faster adjustment to equilibrium when exogenous shocks hit the economy. The world may still have to deal with such shocks in the future, the end of oil reserves, nuclear disaster in the Middle East, the recovery of Japan, or the introduction of cold hydrogen fusion reactors. Whatever lies ahead, entrepreneurs can help deal with it and adjust as long as change opens up opportunities.

Finally, consider the effect of all shocks described above combined. As we argued in section 2.1 that is precisely what happened over the past 25 years. The model then predicts significant wage divergence in the North between the skilled and unskilled and certainly between the entrepreneurs and workers in general, increasing FDI and outsourcing to the South, and a widening of the North–South wage gap. The latter has not been observed in the data as China and India have always been considered part of the South. But taking them out of the sample one sees significant divergence between OECD and non-OECD countries in both wage and income-levels. China and India, on the other hand, are quickly closing the wage and income gap with the North and not surprisingly they receive the bulk of FDI outflow from the North as well. Lower political risk implies that the downward wage pressure in the (rest of) the South is somewhat ameliorated. Our policy experiment of increasing the entrepreneurial capabilities shows that self-interested national policies may yet help to tip the scales and help close the North–South wage and income gaps, while offering new perspectives for countries in the North to remain competitive among their peers in globalizing markets.

2.4 Conclusions

Globalization and technological change have triggered fundamental changes in both the developed and developing countries. This chapter illustrates one mechanism through which these shocks interact. By opening up new opportunities for entrepreneurs new technology and a global market both create a dynamic that shifts the specialization pattern in the global economy. An important implication of our model is that developing countries are gaining the comparative advantage for economic activity in the later stages of the product life cycle, where the classic factors of raw labour and physical capital play an important role. At the same time, globalization has facilitated a shift in comparative advantage in the advanced economies away from mature industries and towards economic activity in the early stage of the product life cycle. This reorientation towards early stage activity was reinforced by the simultaneous rise of a GPT that offers both the means for organizing off-shoring and outsourcing and presented entrepreneurs with many opportunities for new products and services. The coincidence of these trends caused and forced the advanced countries to enter the entrepreneurial economy, whereas it allowed (some) developing countries to rapidly industrialize. Whether the emergence of the entrepreneurial economy is permanent or transitory depends upon the nature of globalization. It does not seem likely, however, based on the model presented here, to expect a return to a mature and industrial global economy any time soon. The political turnaround we have identified as the exogenous shock driving globalization is not likely to be reversed. On the technology side we do know that related new products and services will mature. Under current global competitive conditions it is likely that these mature industries will also quickly transfer routine operations to less expensive regions.

The policy implications of our analysis are thus that developed countries' governments should gear their labour market and social security policies to the new economic reality. Flexibility, mobility, and employability are key assets for workers and self-employed in the entrepreneurial economy. A large and easily accessible low-skilled non-tradables sector such as personal services would have to be developed if welfare states are to survive this pressure.

For the developing world the focus should gradually shift from attracting FDI inflows to building local capabilities and absorbing mature technologies. In the end, the transition to a developed country is possible, but to take off a strategy of gradually moving upstream in the life cycles seems appropriate. Modern growth theory has little to offer for the situation these countries find themselves in. More R&D will not help regions and countries in which not even mature routine industrial activity has been developed. But governments should realize that comparative and competitive advantages are temporary in nature and should not be protected against such dynamics. Stimulating local entrepreneurs to take advantage of the opportunities in global markets is the one policy that can be shown to deliver lasting advantages. By their flexibility, these entrepreneurs can help adapt to possible future shocks as well as the current slump in global economic growth.

Our model does not generate very detailed policy advice for either side. The model's out-of-equilibrium dynamics remain to be studied in detail and its predictions are yet to be tested more rigorously in the data. But we claim at this point that the general direction is clear and further research is warranted to gain a more fundamental understanding of these macroprocesses. We hope to have contributed to such understanding.

APPENDIX

Proof of Proposition 1

The proof follows from assuming this steady-state exists and showing that it constitutes equilibrium in all labour markets. First, consider the markets for production labour. They clear iff the allocation satisfies:

$$
\begin{aligned}
L^N_{H*} &= n^N_H \alpha^{\frac{1}{1-\alpha}} b^{\frac{\alpha}{1-\alpha}} EP^{\frac{\alpha}{1-\alpha}} (w^N_H)^{\frac{1}{\alpha-1}} \\
L^N_{L*} &= n^N_L b^{-1} EP^{\frac{\alpha}{1-\alpha}} \left(\min\left[\frac{w^N_H}{b}, \frac{w^N_L}{\alpha b}\right]\right)^{\frac{1}{\alpha-1}} = n^N_L b^{\frac{\alpha}{1-\alpha}} EP^{\frac{\alpha}{1-\alpha}} \left(\min\left[\frac{w^N_H}{w^N_L}, \frac{1}{\alpha}\right]\right)^{\frac{1}{\alpha-1}} (w^N_L)^{\frac{1}{\alpha-1}} \\
L^N_{H*} &= n^S_L b^{-1} EP^{\frac{\alpha}{1-\alpha}} \left(\min\left[\frac{w^N_L}{b}, \frac{w^S_L}{\alpha b}\right]\right)^{\frac{1}{\alpha-1}} = n^S_L b^{\frac{\alpha}{1-\alpha}} EP^{\frac{\alpha}{1-\alpha}} \left(\min\left[\frac{w^N_L}{w^S_L}, \frac{1}{\alpha}\right]\right)^{\frac{1}{\alpha-1}} (w^S_L)^{\frac{1}{\alpha-1}}
\end{aligned}
\tag{A1}
$$

Taking growth rates on both sides implies:

$$
\begin{aligned}
0 &= \frac{\dot{n}^N_H}{n^N_H} + \frac{\dot{E}}{E} + \frac{\alpha}{1-\alpha}\frac{\dot{P}}{P} + \frac{1}{\alpha-1}\frac{\dot{w}^N_H}{w^N_H} \\
0 &= \frac{\dot{n}^N_L}{n^N_L} + \frac{\dot{E}}{E} + \frac{\alpha}{1-\alpha}\frac{\dot{P}}{P} + \frac{1}{\alpha-1}\frac{\dot{w}^N_L}{w^N_L} \\
0 &= \frac{\dot{n}^S_L}{n^S_L} + \frac{\dot{E}}{E} + \frac{\alpha}{1-\alpha}\frac{\dot{P}}{P} + \frac{1}{\alpha-1}\frac{\dot{w}^S_L}{w^S_L} \\
&\frac{\dot{w}^S_L}{w^S_L} = \frac{\dot{w}^N_L}{w^N_L} = \frac{\dot{w}^N_H}{w^N_H}
\end{aligned}
\tag{A2}
$$

where the latter constraints follow from the fact that relative wages appear in the expressions. From these conditions we can derive that in any steady-state equilibrium wages must grow at the same rate and hence so must all sectors. This in turn implies in (15) that the number of firms in each sector must grow at the same constant rate as n^A.

Now consider the market for entrepreneurial talent. It clears when (17) holds. Under the conditions derived above, the reward to entrepreneurship can be shown to grow at rate:

$$\frac{\dot{w}_T}{w_T} = \frac{\dot{n}^A}{n^A} + \frac{\dot{v}^N_H}{v^N_H} = \frac{\dot{n}^A}{n^A} + \frac{\dot{\pi}^N_H}{\pi^N_H} = \frac{\dot{n}^A}{n^A} + \frac{\dot{E}}{E} - \frac{\dot{n}^N_H}{\dot{n}^N_H} = \frac{\dot{E}}{E} \tag{A3}$$

This implies that, if relative wages are to be constant, all must grow at the same rate as total expenditures and (A2) reduces to:

$$0 = g + \frac{\alpha}{1-\alpha}\frac{\dot{P}}{P} + \frac{\alpha}{\alpha-1}\frac{\dot{E}}{E} \tag{A4}$$

If we then normalize expenditure to 1 growth rates of profits and consequently innovation values reduce to $-g$ and the interest rate is endogenously determined by (2) to equal ρ. There is therefore a steady-state equilibrium that is fully characterized by:

$$\begin{aligned}
&\frac{\dot{n}^A}{n^A} = \frac{\dot{n}^N_H}{n^N_H} = \frac{\dot{n}^N_L}{n^N_L} = \frac{\dot{n}^S_L}{n^S_L} = -\frac{\dot{\pi}^N_H}{\pi^N_H} = -\frac{\dot{\pi}^N_L}{\pi^N_L} = -\frac{\dot{\pi}^S_L}{\pi^S_L} = -\frac{\dot{v}^N_H}{v^N_H} = -\frac{\dot{v}^N_L}{v^N_L} = -\frac{\dot{v}^S_L}{v^S_L} = g\\
&\frac{\dot{E}}{E} = \frac{\dot{w}^N_H}{w^N_H} = \frac{\dot{w}^N_L}{w^N_L} = \frac{\dot{w}^S_L}{w^S_L} = \frac{\dot{w}_T}{w_T} = r - \rho = 0\\
&\frac{\dot{P}}{P} = \frac{\alpha-1}{\alpha}g
\end{aligned} \tag{A5}$$

Q.E.D.

Proof of Proposition 2

Consider being out of equilibrium such that $n^N_H/n^N_L>(n^N_H/n^N_L)^{SS}$. Then by Table 2.1 $\pi^N_H/\pi^N_L<(\pi^N_H/\pi^N_L)^{SS}$ and consequently by (12)–(13) $v^N_H/v^N_L<(v^N_H/v^N_L)^{SS}$ which implies that $T_I<T_I^{SS}$ and $T_D>T_D^{SS}$ such that $\dot{n}^N_H/n^N_H < g$ and $\dot{n}^N_L/n^N_L < g$. Therefore we move towards the equilibrium as (15) then implies that n^N_H/n^N_L falls towards $(n^N_H/n^N_L)^{SS}$. That process will end in the steady-state. A similar argument can be made for all possible deviations from the steady-state allocation of entrepreneurial activity, which proves that the steady-state is both unique and stable.

Q.E.D.

Notes

1. Outsourcing refers to placing parts of the production chain out of the firm and does not necessarily cross international borders. Off-shoring refers to placing (parts of) the firm's activities abroad but maintaining control.
2. Longer time series show a stagnant period in the Interbellum (1918–39) and during the First and Second World Wars for well-known and obvious reasons. The export share in 1913 was 8.7 per cent and stood at 7 per cent of global GDP in 1950.
3. Antras (2005) offers an incomplete contract model to explain why Northern firms and entrepreneurs, rather than Southern imitators, are the main initiator and vehicle of international technology transfer.
4. 'Markets Go Global', *Economist*, 20 September 1997.
5. See for example Grossman and Helpman (1991).
6. Note that in fact we assume that products only mature as the incumbent producer engages in such R&D. It is not an automatic process as for example in Krugman (1979) or Arrow (1962).
7. Here we have to assume they cannot do so as long as the firm employs them. Some non-disclosure agreements typically ensure that they do not infringe upon the incumbent's profits while and even some time after being employed there.
8. In reality there are all kinds of competitive advantages. The Northern incumbent may enjoy lower transportation costs to the largest markets, protective tariffs and subsidies, complementary sectors, etc.
9. Obviously it is worthwhile only if the discounted present value of this profit flow exceeds the costs of making the required innovation.
10. In fact we assume their cost advantage and tacit knowledge give them a sufficient edge, so we see only incumbent firms doing product development.
11. This is our parameter for political risk.
12. Results do not change qualitatively and become stronger if we allow for entrepreneurs to not compensate incumbents. Obviously that makes outsourcing even more attractive.
13. Alternatively one could think of R&D labour, engineering talent, or organizational and managerial talents. All that is required to justify this assumption is that all innovative activities require the same labour that is a different type of labour than that used in production.
14. See Sanders (2005) for more details on how this knowledge parameter can be endogenized. In standard new growth models it is frequently assumed to be equal to n, accumulated innovation. In this model this would also introduce endogenous growth but prevent us from experimenting with shocks to the knowledge stock. The introduction of a new GPT, however, is exactly that, a positive shock to n^A.
15. One could consider setting w_T equal to the high-skilled wage in the North as it arguably presents the opportunity costs for the entrepreneur. However, this complicates the model considerably without adding much in terms of insight in the mechanisms we are primarily interested in. Sanders (2005) shows the equivalence in qualitative results in a different (two stage life cycle) but closely related model.

References

Acemoglu, D. (2002a) 'Technical Change, Inequality and the Labour Market', *Journal of Economic Literature*, 40 (1): 7–72.

———. (2002b) 'Directed Technical Change', *Review of Economic Studies*, 69 (241): 781–809.

Acs, Z., and D. B. Audretsch (1990) *Innovation and Small Firms*. Cambridge; MA: MIT Press.

———. (1993) *Small Firms and Entrepreneurship: An East-West Perspective*. Cambridge: Cambridge University Press.

Antras, P. (2005) 'Property Rights and the International Organization of Production', *The American Economic Review*, 95 (2): 25–32.

Arrow, K. (1962) 'The Economic Implications of Learning by Doing', *Review of Economic Studies*, 29 (3): 155–73.

Audretsch, D. B. (2007) *The Rise of the Entrepreneurial Economy*. Berlin: Springer Verlag.

Audretsch, D. B., and Z. Acs (eds) (2005) *Handbook of Entrepreneurship Research: An Interdisciplinary Survey and Introduction*, Vol. 1. Berlin: Springer Verlag.

Audretsch, D. B., M. A. Carree, A. J. van Steel, and A. R. Thurik (2002) 'Impeded Industrial Restructuring: The Growth Penalty', *Kyklos*, 55 (1): 81–98.

Autor, D. H., L. F. Katz, and M. S. Kearney (2005) 'Trends in US Wage Inequality: Re-Assessing the Revisionists', Working Paper 11627. Cambridge, MA: National Bureau of Economic Research.

Birch, D. (1981) 'Who Creates Jobs?', *The Public Interest*, 65 (Fall): 3–14.

Bresnahan, T. F., and M. Trajtenberg (1995) 'General Purpose Technologies "Engines of Growth"?', *Journal of Econometrics*, 65 (1): 83–108.

Brock, W. A., and D. S. Evans (1989) 'Small Business Economics', *Small Business Economics*, 1 (1): 7–20.

Card, D., and J. E. DiNardo (2002) 'Skill-Biased Technological Change and Rising Wage Inequality: Some Problems and Puzzles', *Journal of Labour Economics*, 20 (4): 733–83.

Chandler, A. (1990) *Scale and Scope: The Dynamics of Industrial Capitalism*. Cambridge, MA: Harvard University Press.

Davis, S. J., J. Haltiwanger, and S. Schuh (1996a) *Job Creation and Destruction*. Cambridge, MA: MIT Press.

———. (1996b) 'Small Business and Job Creation: Dissecting the Myth and Reassessing the Facts', *Small Business Economics*, 8 (4): 297–315.

EIM (2002) 'SMEs in Europe. The European Observatory for SMEs', Vol. 2. Report submitted to the European Commission, Enterprise Directorate-General. KPMG Special Services and EIM Business School & Policy Research, Düsseldorf.

Feenstra, R. C., and G. H. Hanson (1996) 'Globalization, Outsourcing and Wage Inequality', *The American Economic Review*, 86 (2): 240–5.

———. (1999) 'The Impact of Outsourcing and High-Technology Capital on Wages: Estimates for the United States, 1979–1990', *The Quarterly Journal of Economics*, 114 (3): 907–40.

———. (2004) 'Global Production Sharing and Rising Inequality: A Survey of Trade and Wages' In J. Harrigan and E. K. Choi (eds), *Handbook of International Trade*. London: Basil Blackwell.

Florida, R. (2003) *The Rise of the Creative Class*. New York: The Perseus Book Group.

Goldin, C., and L. F. Katz (2001) 'Decreasing (and then Increasing) Inequality in America: A Tale of Two Half-Centuries'. In F. Welch (ed.), *The Causes and Consequences of Increasing Income Inequality*. Chicago, IL: University of Chicago Press.

Gottschalk, P., and R. Moffit (1994) 'The Growth in Earnings Instability in the U.S. Labour Market', *Brookings Papers on Economic Activity*, 2: 217–72.

Greenwood, J. (1997) *The Third Industrial Revolution: Technology, Productivity and Income Inequality*, AEI Studies on Understanding Economic Inequality. Washington, DC: AEI Press.

Greenwood, J., and G. Uysal (2004) 'New Goods and the Transition to a New Economy', Working Paper 10793. Cambridge, MA: National Bureau of Economic Research.

Grossman, G. M., and E. Helpman (1991) *Innovation and Growth in the Global Economy*. Cambridge, MA: MIT Press.

Helpman, E. (1998) *General Purpose Technologies and Economic Growth*. Cambridge, MA: MIT Press.

Iversen, T., and F. Rosenbluth (2006) 'The Political Economy of Gender: Explaining Cross-National Variation in the Gender Division of Labour and the Gender Voting Gap', *American Journal of Political Science*, 50 (1): 1–19.

Juhn, C., K. M. Murphy, and B. Pierce (1993) 'Wage Inequality and the Rise in Returns to Skill', *Journal of Political Economy*, 101 (3): 410–42.

Katz, L. F., and D. H. Autor (1999) 'Changes in the Wage Structure and Earnings Inequality'. In O. Ashenfelter and D. Card (eds), *Handbook of Labour Economics*, Vol. 3. North Holland: Elsevier.

Keller, W. (1996) 'Absorptive Capacity: On the Creation and Acquisition of Technology in Development', *Journal of Development Economics*, 49: 199–227.

Kortum, S., and J. Lerner (1997) 'Stronger Protection or Technological Revolution: What is Behind the Recent Surge in Patenting?', Working Papers 6204. Cambridge, MA: National Bureau of Economic Research.

Krugman, P. (1979) 'A Model of Innovation, Technology Transfer and the World Distribution of Income', *Journal of Political Economy*, 87 (2): 253–66.

Lane, P., and G.-M. Milesi-Ferretti (2003) 'International Financial Integration', Discussion Paper 3769. London: Centre for Economic Policy Research.

Loveman, G., and W. Sengenberger (1991) 'The Re-Emergence of Small-Scale Production: an International Comparison', *Small Business Economics*, 3 (1): 1–37.

Machin, S., and J. van Reenen (1998) 'Technology and Changes in Skill Structure: Evidence from Seven OECD Countries', *Journal of Quarterly Economics*, 113 (4): 1215–244.

Mayer, J., and A. Wood (1999) 'South Asia's Export Structure in a Comparative Perspective', Working Paper 91. Sussex: Institute of Development Studies.

Nickell, S., and B. Bell (1996) 'Changes in the Distribution of Wages and Unemployment in OECD Countries', *The American Economic Review*, 85 (2): 302–8.

OECD (2005) *Micro-Policies for Growth and Productivity: Final Report*. Paris: OECD.

Sanders, M. (2005) *Technology and the Decline of Unskilled Labour Demand: A Theoretical Analysis of the US and European Labour Markets*. Cheltenham: Edward Elgar.
Sanders, M., and B. Ter Weel (2000) 'Skill-Biased Technical Change: Theoretical Concepts, Empirical Problems and a Survey of the Evidence', Working Papers 00–8. Aalborg: DRUID.
Schultz, T. W. (1961) 'Investment in Human Capital', *American Economic Review*, 5 (1): 1–17.
Torres, R. (2001) *Towards a Socially Sustainable World Economy: An Analysis of the Social Pillars of Globalisation*. Geneva: International Labour Office.
UNCTAD (2003) *World Investment Report*. New York and Geneva: United Nations Conference on Trade and Development.
Vernon, R. (1966) 'International Investment and International Trade in the Product Cycle', *Quarterly Journal of Economics*, 80 (2): 190–207.
WTO (2005) *International Trade Statistics*. Geneva: World Trade Organization.

3

Firm Ownership and Entrepreneurship

Alice H. Amsden

3.1 Introduction

In perfectly competitive markets, the nationality of a firm's owner does not matter for economic development and entrepreneurship. A foreign-owned enterprise (FOE) from a developed country or a privately owned enterprise (POE) from a developing country are equally capable if both have access to the same inputs and marketing opportunities (ignoring state-owned enterprises (SOEs) and tiny enterprises). Where the theory of free competition reigns, developing countries should open their arms to investments from all types of enterprises in order to maximize jobs, as many Asian countries have done, using export processing zones. Ownership, measured by votes of shareholders or boards of directors, is immaterial to performance.[1]

Matters change drastically, though, when competition depends on monopolistic assets and market theory no longer rigorously holds. This happens specifically in 'mid-tech' industries such as shipbuilding, steel, and heavy machinery, and 'mature' high-tech industries, such as manufacture of calculators, computers, and cell phones, after millions of units from advanced countries have already been sold but demand is still booming. Knowledge, brand names, political clout, and other sources of market power vary by firm and influence market outcomes. Then, ownership matters. It matters exponentially depending on how diversification from mid-tech into high-tech occurs—by an existing firm or by a new entrant. Given that multinational firms tend to operate within a single product area, however broadly defined, if diversification from mid-tech to high-tech is left to the multinationals, then each diversification will tend to involve a new foreign entrant, which may be an expensive way for a developing country to flesh out its industrial base (including mining, agro-industry, and services).

How then should we think about firm ownership and entrepreneurship in a developing context? Will the capabilities to enter 'new' (for the developing country) industries be more entrepreneurial (measured by project execution skills) or will the multinational firm be better? And once entry occurs, can we expect the multinational subsidiary to be innovative if it exists at the end of a long chain of command that originates in large corporate headquarters overseas? Given that the well-managed nationally owned enterprise knows its own environment better than does the multinational, and is on its own (except for government support), without help from a large corporate family, will the need to survive drive the domestically owned entity to be more entrepreneurial?

The relationship between ownership and entrepreneurship is, thus, intimately related, and raises many more questions than this chapter can answer. Here, we selectively try to argue, not always with empirical support, that ownership matters outside the most labour-intensive industries, because FOEs and POEs differ in degree of entrepreneurship and bureaucracy.

Because FOEs have controlled more competitive assets than POEs since the First Industrial Revolution, they have gained the upper hand and the respect of the elite business schools that tend to view them as the best policy choice for backward countries to follow. FOEs are highly productive, the argument runs, and somehow transfer knowledge (measured by 'spillovers') to local enterprise, so invite them into your midst.

But this outlook has turned out to be extremely short-run. FOEs can 'crowd out' POEs. They can break their back before they have a chance to acquire their own assets, and the acquisition of their own assets may be far better for economic development than FDI, or joint ventures. 'Foreign firms' do pass on important information to newcomers, but not through FDI. Foreign firms that are teachers tend to be vendors of parts and components located in third countries. They help by giving all their customers a roadmap of their industry. FOEs in direct competition with POEs are not necessary for economic development to flourish, and it is dangerous for a promising POE to confront a privileged FOE in its own back yard, often with the backing of the FOE's powerful government (a role Washington played in the foreign dominated Latin American automobile industry, for example).

Because assets differ systematically between FOEs and POEs in their respective stages of evolution, we would argue that FOEs may not contribute more to economic development in monopolistic industries than POEs. For one, they cannot create the highest managerial talents, which are always kept at home in corporate headquarters. The skills and salaries of the chief executive officer (CEO) and other top managers are never located abroad. Nor can FOEs 'globalize' a developing country in the form of outward foreign investment.

In order for a developing country to invest abroad, it must have its own companies.

I would argue that the best of POEs in the fastest growing emerging economies (e.g. Korea's Samsung, India's Tata, and Brazil's Embraer) tend to be more entrepreneurial than FOEs. FOEs today are bureaucratic—they operate with relatively dense levels of management and continue to cookie-cut a single model throughout the world that has proved to be highly lucrative for them but not necessarily for emerging economies. Bureaucracy can be measured by a number of signatures, review boards, management layers, or people thrown at a problem. For now, during a growth phase when most POEs enjoy both family ownership and professional management, they display a highly dynamic entrepreneurship. Akin to what Schumpeter said, they formulate novel ideas (locally), coordinate resources, enter and exit 'new' (for them) industries at lightning speed (a highly profitable skill unmentioned by Schumpeter), and restructure industries in a creative way that in labour abundant countries do not involve the mass lay-offs typical of American retrenchments (China's restructuring, say, of the First Automobile Works, involved only modest redundancies).

The thin layer of bureaucracy in POEs improves information flows. With fast transmissions of information, POEs have perfected the asset of short time to market, especially in Asia. They are superquick in picking new industries to enter and then designing the integration of parts and components to win the global race to the market. In the case of one POE in the Indian pharmaceutical industry, it reached the market faster than the Indian subsidiary of the multinational that had invented the drug in the first place (Mourshed 1999). POEs, not FOEs, should take credit for industrialization in the developing world. Industry by industry, they diversified forcefully and fast. In the process, they themselves became highly diversified as business groups.

3.2 Business groups

POEs have tended to take the form of diversified business groups because, unlike specialized firms in advanced countries with revolutionary technology, POEs from developing countries are still competing without worldclass products and know-how. It is too risky for them to specialize in a single technology family. In almost every market, they are behind the technological frontier and must move from industry to industry as opportunities arise. To do so, POEs must develop alternative skills, not just to do battle with FOEs but also to compete against other POEs from developing countries, all of which depend on low wages to compete. Long hours and voluntary or involuntary overtime help developing countries reach consumers at record speed, but only skills, not hours of work, differentiate the best.

The diversification process itself holds the key to group success. The more groups diversify, the greater their experience and acquired skills at diversification (or 'project execution', from buying technology to starting-up operations; see Amsden and Hikino 1994). This allows them to get a headstart over other entrants into the 'new' industries (and services) that their governments open or that they capture from overseas. Groups can enter these 'new' industries with great speed and low cost aided by project execution skills, in a virtuous circle.

The greater the scope of their operations are, then the greater the visibility of a group, and the greater its ability to attract the best talent to reach world norms of productivity at the industry-wide level. This talent brings the diversified firm the production know-how that specific industries require. Expertise in many industries gives groups an advantage with large, multiproduct vendors that provide them with further knowledge and better industry-level capabilities.

The swiftness of POEs, which raises profitability, stems partly from their membership in a business group that is managed by a single owner, who knows where resources lie and who can transfer them from one subsidiary to another. The fungibility of resources among the subsidiaries of a single group, especially capital ('cross-investments'), makes it difficult for FOEs to engineer takeovers of POEs, which is why they are disliked on Wall Street (Lawrence 1993). Instead of searching for the right person for a job on the market, which takes time in the developing world, the group can find a cachet internally from one or another subsidiary. Groups even harbour a captive market for individual subsidiaries. The employees of Korea's groups bought their life insurance and their automobiles from sister subsidiaries.

Groups benefit from the inter-sectoral learning made possible by the absence of operating in a single market (Dosi 1988). Paint problems in automobile making can benefit from the experience of solving paint problems in the shipbuilding industry—both within the scope of a group such as Korea's Hyundai. A business group in the developing world today is typically in its first, second, or third generation of family ownership (a 60-year cycle roughly). It is free from bureaucracy at the top while enjoying professional management at the bottom and middle. Joint ventures are regarded in much of the business school literature as a compromise between FOE and POE, good for both, but the protracted negotiations that joint ventures typically involve slow decision-making. Decision-making is far faster in a diversified group, or an alliance of national domestic groups, than in a foreign joint venture: compare the speed of entry and operations in China's automobile industry of Cheri, a national champion, and joint ventures involving General Motors and Volkswagen. Both Malaysia and China, once heavily dependent on FDI, have begun to reconsider their strategy given what they consider to be inadequate technology transfer by the multinational firm.

POEs tend to have more knowledge of the local business environment than FOEs. Charles Kindleberger argued as early as 1970 that proximity creates advantages for the national investor. But he was writing before the mass migration of Third World managers and engineers to study, live, and work in advanced countries, permanently or temporarily. This coming and going between continents has given leading POEs two types of knowledge, not just one. The people who know their own native environment also know the richest economies abroad. Managers born in India or Brazil, for example, who then study or work in the United States, acquire a unique take on globalism that Americans, knowing only one culture and one language, do not have.

Diversified business groups have excelled especially in mature high-tech industries, or industries with advanced technology whose demand is still growing but whose profitability has begun to fall. When the Radio Corporation of America invested in Taiwan and Mexico to manufacture TVs in the 1950s, it opened its own subsidiaries. When NOKIA goes to Taiwan today to lower the production costs of its cell phones, it outsources production to Taiwan-owned companies. Taiwan's POEs have their own skill sets, and ramp up quickly to achieve volume in the face of falling profit rates, not least of all by creating internal construction and automation arms to hasten the conversion of old factories (e.g. in textiles) to new ones (in cell phones). Taiwan's POEs excel in integrating the hundreds of parts and components bought outside that comprise, say, a notebook, given their skills in modularity and design (and weakness in designing a single advanced component). Within hours or days rather than weeks, a product made by a POE can reach a FOE, making the wholesale relocation of manufacturing activity to developing countries all the more likely.

This is the gist of the argument that crowding out of POEs by FOEs in monopolistic industries (mid-tech and mature high-tech) is not development friendly. To celebrate FOEs is to close one's eyes to the knowledge-based assets of POEs. Now that POEs are growing, their speed to market, project execution skills, and capabilities in assembly make it easier to answer the question of whether or not they can ultimately reach the world technological frontier. Why not, given their assets?

Below, we try to answer the question of why POEs are more evident in some emerging economies than in others. The answers possibly lurk in a few facts of colonialism and decolonization, one of the great events of the twentieth century.

3.3 Decolonization

Some say colonialism was indispensable in enlightening backward people (Ferguson 2003); others say it retarded progress once colonial peoples understood foreign ways. Pre-Second World War manufacturing experience arose in only a few colonies, so as a system of development, colonialism clearly had its limits. Among countries with pre-war manufacturing experience, only a few succeeded after decolonization in freeing themselves not only from foreign rule but also from foreign firms. Those that succeeded in freeing their industries from FOEs became the most successful in the developing world after 1950, another slap in the face to colonialism.

POEs were a scarce institution in the colonial world. Manufacturing experience, involving not just physical dexterity but the embodiment of that dexterity in a 'firm'—a complex social entity with knowledge of accounting, finance, labour management, and marketing—was the key to economic development but a scarce competitive asset. No developing country after the Second World War entered the orbit of modern industry without pre-war manufacturing experience, measured by the share of a country's manufacturing in GDP (excluding countries where manufacturing comprised only a single industry, as in the textile industry of Egypt and Pakistan, the oil industry of Venezuela, and the processed foods industry of the Philippines). Only one dozen developing countries managed to acquire diverse pre-war manufacturing experience, and POEs arose only in these countries (Argentina, Brazil, Chile, China, India, Indonesia, Korea, Malaysia, Mexico, Taiwan, Thailand, and Turkey; see Amsden 2001, 2007).

Manufacturing experience derived from three sources: pre-modern know-how that pre-dated Western influence (in China, India, Mexico's woollen industry, and possibly Ottoman Turkey); émigrés, who brought business acumen with them to Latin America (from Europe and the US) and the Pacific Basin (from China); and colonial companies, which built industries in their own conquered territories. Japan's coprosperity sphere was a hub. Korea was industrialized to act as a bridgehead to Manchuria. Taiwan was belatedly industrialized to help Japan conquer South East Asia. As Japan's war drums beat louder, the UK introduced defensive industry in Malaysia, and the Dutch and Americans did the same in Indonesia and the Philippines respectively. Fascist governments in both Japan and Thailand in the 1930s collaborated to build military-related industries.

Latin American émigrés were probably the most successful in transferring know-how, because Latin America had the highest per capita income in the developing world after the Second World War. But post-war decolonization wrought radical change.

Decolonization did not affect Latin America at all because Latin America had gained its independence in the early nineteenth century. After the Second World War, the same powerful political groups and the same FOEs survived as before (Pirelli was the first multinational to arrive in Argentina in 1917, and still exists). This inertia left Latin America with relatively little space for POEs. Industrial leadership was lethargic until a developmental state arose. Only then did Brazil, for example, create Embraer, Petrobras, and other SOEs.

Starting with India in 1947, colonies slowly got their independence. The first African colony to cast off colonial ties was Ghana in 1957, and the last was South Africa in 1989. But the countries that became fertile crescents for POEs, mostly in Asia, not only shed colonial rule but also kicked out colonial companies. China expropriated them. India drove them out by fear or competition. Korea and Taiwan acquired Japanese properties, especially subsidiaries of big banks. Indonesia inherited around 400 Dutch companies. Malaysia bought British properties on the London Stock Exchange (Amsden 2001, 2007).

In one way or another, these former colonies 'crowded out' foreign firms and created space for their own national POEs to flourish. Those that did not create this space—the Philippines, for example, got rid of American foreign rule but was stuck with uninventive American companies—performed badly.

Virtually all POEs took the form of diversified groups, but there is disagreement about why this business model arose in such a wide range of poor countries. Disagreement has tended to be narrow and one-sided, especially if groups are analysed as market phenomena, without the goal of economic development directly in mind (Khanna and Yafeh 2007).

To let in some fresh air, it is helpful to think of groups as either money machines, that mobilize finance for development (the market approach), or as institutions that build knowledge, skills, and technological capabilities (the institutional approach). Some say groups arose to pool family savings when financial markets were weak. But this assumes that savings were scarce rather than 'shy' (savers awaited the appearance of firms with profitable skills to invest in). Others argue that to stay independent from the sorties of FOEs, the group form emerged—as it had done in Japan much earlier—and empowered itself by building knowledge-based assets.

If a country runs a persistent surplus in its balance of payments such that savings exceed investment, as in India from 1835 to 1946, then capital shyness may be said to exist (Das 1962; Banerjee 1963; Maddison 1971; Bagchi 1972). There was no scarcity of investment capital in India, however poor; it was exporting capital. Alternatively, shyness can mean the excess product that a country either does or can produce above subsistence (Riskin 1975), as in pre-industrial China. As early as the 1840s, there was substantial qualitative evidence that when profitable opportunities in India and China arose, capital came out of the woodwork, and that the family form of business is not tightly tied to how capital is raised.

In the case of India: the early textile mills were not exceptionally costly ventures by local standards. A company could get into operation in Bombay for an investment of Rs. 500,000 to Rs. 1 million or about £50,000 to £100,000 at prevailing exchange rates. This covered cost of land, buildings, equipment, and inventory. Many other types of enterprise projected in the same period involved sums as great or greater. Shares were issued in units of Rs. 2,500 or, more typically, Rs. 5,000. These were not amounts intended to attract the small investor. Yet the number of people in Bombay with sums to risk in promising enterprises was sufficiently great so that when the Oriental mill was floated in 1854 with paid-up capital of Rs. 1,250,000 divided into 500 shares of Rs. 2,500, no one was permitted to subscribe for more than four shares (Morris 1983).

According to a partner of Tata Sons and Company, India's largest group: 'The public in India, especially in Bombay, are ever ready to put their money in mill concerns started by individuals or firms who have a reputation for honesty and efficiency, and who have a good deal of mill experience' (Chandavarkar 1994: 67). With a successful business in textiles, Tata was able to raise money from the private sector in 1907 for a large steel mill. The total capital of the new company was subscribed by the Indian public in a remarkably short space of a few weeks, the number of shareholders being about 7,000 (Fraser 1919).

China's potential economic surplus in 1933 was also estimated (by Riskin 1975) to be very large, possibly more than 25 per cent of gross national product. Given the operations of 'compradors' (merchants to foreign business) 'contrary to the generally held view, an important reason for China's relatively slow economic development in the nineteenth century was not the scarcity of capital, because large amounts of Chinese funds were readily available' (Hao 1986: 348). The Chee Hsin Cement Company, for example, allegedly had no trouble raising capital after the turn of the century. Its 29 shareholders held diversified portfolios, with interests in other industries, commerce, and banking (Feuerwerker 1967).

Rather than being dirt poor, the problem was high risks from low skills. In India, only five of the pioneering textile firms of the nineteenth century survived until the First World War. Out of a total of 97 mills erected in Bombay between 1855 and 1925, 12 were burnt down or else were closed and dismantled, 16 transferred their managing agencies voluntarily, and 45 went into liquidation and were reconstructed under other names (Rutnagur 1927). There occurred a very large number of company failures, with the result that 'the Indian investor, habitually shy, became shyer still' (Das 1962: 162).

Causality thus seems to run from 'no skills' to 'no money' rather than the reverse, both before and after decolonization. Shyness of capital seems to characterize much of Africa today, according to studies of Canada's North–South Center in Ottawa.

There was huge speculation before the Second World War, even in Japan, but in commodities such as cotton and silk, not financial instruments. After the Second World War, the stock markets in India boomed. But this only goes to show that groups could get the capital they wanted without family strings attached.

Skill-intensity differentiated groups in developing countries from conglomerates in the US, which regarded their affiliates as an asset to buy and sell. The business groups of the developing world, by contrast, regarded their affiliates as long-term commitments, to be nursed to health rather than put up for adoption if sick. Few affiliates until very recently could be bought or sold.

Thus, most group characteristics do not fit the market paradigm, but have proved to be development friendly despite an enormous amount of angry criticism of their structure and strong government ties. The developing countries where POEs became strongest were the developing countries with the longest history of manufacturing experience and not trade or free markets. Decolonization was most developmental when it included the expulsion of FOEs.

3.4 Services

As manufacturing migrated in ever increasing numbers to developing countries, the FOE found a sanctuary in services. It had a helping hand from the Great Powers, ranging from the US Treasury, which pressured emerging economies to open their markets and allow foreign takeovers, to the most expensive business schools, which supplied the rationales for doing so and the WTO, which rewrote the regulatory rules for privatizing government-owned services and operating private services along Western accounting and financial lines. How did POEs fare, when their own presence in the service sector of FOEs was negligible, a hindrance to their imitative learning?

Three points about national ownership in the service sector emerge from Taiwan. First, the Taiwanese government promoted national ownership in services as it had done in manufacturing, despite global cries for laissez faire. It bet on the speed of learning and skill formation of its groups to keep services in national hands. It won from the US government a five-year reprieve before some services were opened to FOEs. This gave them time to speed up learning.

Second, the big national players in services became mainly old business groups, not specialized national service providers. (The top ten companies in retail, wholesale, and department stores, the largest service sub-sector, were all groups.) Third, many groups transferred know-how to services from traditional industries (Amsden and Chu 2003).

In the case of cell phone services, all groups built up their operating systems within a year, and then, three months later, entered into fierce price competition with each other. 'Ramp ups and start ups were extremely fast by world standards', an advantage gained from older industries (Amsden and Chu 2003: 137). In sectors where financial assets were lacking, local firms allied with each other. Foreign joint ventures grew, but alliances were viewed as a faster route to market entry, especially in financial services and telecommunications.

A company in the textile and construction industry, Ruentex, founded RT-Mart International, along the lines of Costco, the American mass wholesale warehouse. After it opened its first store and bought two others, Ruentex quickly ramped up to a total of 16 stores to acquire brandname recognition and good vendor service. A year later it opened its first outlet in China, which soon expanded to eleven outlets in all. Speed came from Ruentex's traditional construction and development arm, which designed and built Ruentex's retail outlets. Capital to finance diversification into services came from Ruentex's old textile business, which had also invested in a financial securities company and then, the first global Taiwan fund. Once in finance, Ruentex formed a partnership with Aetna Insurance and then bought shares in the Sinopec Bank.

The old economy thus became a bridgehead to services, given the suppleness of business groups. Due to differences in skills, some services became completely foreign dominated (advertising) while others fell mostly under local control (fast foods). Overall, POEs kept their grip on the local economy, which maintained steady growth and low unemployment at a time when manufacturing was migrating en masse to the Mainland. As Taiwan's service sector built up capacity in China, service businesses in Taiwan were strengthened. Just as national capital in manufacturing helped entry into services, so national capital in services helped entry into the Chinese market. Around 60 per cent of total foreign investment in China comes from overseas Chinese POEs.

3.5 Conclusion

Outside perfectly competitive industries, FOEs and POEs are different animals, at least when POEs are still in their first, second, or third generation of family ownership. The POE is entrepreneurial and the FOE tends to be bureaucratic (an assertion we have not dealt with in this chapter). The evidence favours the POE rather than the FOE as the agent of industrialization in the developing world, diversifying from the simplest industries to the most complex, on the basis of latecomer types of skills, especially related to project execution and speed to market.

POEs became strongest in developing countries with pre-war manufacturing experience, whose decolonization involved not just kicking out foreign rulers but also kicking out foreign firms (China, India, Korea, Taiwan, Malaysia, Indonesia, etc.). Space was made for domestically owned POEs to spread their wings, thereby crowding out FOEs and creating the opportunities for developing countries to reach the highest levels of skills and salaries, and to expand overseas. Countries can globalize only on the basis of nationally owned companies. Where the crowding out of FOEs failed to happen, as in the Philippines and Brazil (which never experienced post-war decolonization and where multinationals stayed put), POEs were weaker as an agent of growth.

The diversification pattern of national economies was mirrored in the diversification pattern of business groups, the most popular form of POE. As groups repeatedly diversified, they became good at diversifying—that is, they mastered the skills bundled in project execution—ranging from acquiring technology to starting up operations. With few levels of bureaucracy, the group's most talented professional engineers were known to top management, and could be mobilized for new ventures quickly, rather than sought on the market. With speed and know-how in starting up new operations, groups became extraordinarily fast to market, from ramping up new facilities to integrating parts and components. Today, countries such as Korea and Taiwan can deliver different varieties of notebooks and cell phones to world markets in a matter of hours. Speed is an indigenous asset.

All developing countries after the Second World War were capital scarce, which suggests that the group form of business emerged to mobilize capital on a family basis. But in many countries like India and China (and in many African countries today), capital was more shy than scarce. Capital came out of the woodwork as groups acquired know-how and lowered perceived risk. This attracted the capital of non-family members. The rise of the group form of business in the developing world is best understood in relation to knowledge-based assets, not finance.

In former colonies that ejected foreign firms and at the same time built new entrepreneurial enterprises, globalization began to take the form of brain drain, and then reverse brain drain. POEs soon became blessed with two types of knowledge compared with FOEs: they knew intimately their own native environment and language, and they knew the environments of FOEs, where they studied, worked, and happily lived, and learned English. This contrasts with the limited learning of the US and Japan, which know intimately only their own language and way of doing business. The deep, dual knowledge of the POE argues in favour of its long-run success.

A third type of animal is the joint venture between FOE and POE, and it appears as an attractive mutant for economic development. But joint ventures tend to have protracted negotiations between partners and to move slowly (compare the entry of Cheri on the one hand and General Motors and Volkswagen on the other into the Chinese automobile industry). Nor is history in their favour. After all, no joint venture is known to have emerged between David and Goliath.

Note

1. For further details, see Amsden 2009.

References

Amsden, A. H. (2001) *The Rise of 'the Rest': Challenges to the West from Late-Industrializing Economies*. New York: Oxford University Press.

——. (2007) *Escape from Empire: The Developing World's Journey Through Heaven and Hell*. Cambridge, MA: MIT Press.

——. (2009) 'Nationality of Firm Ownership in Developing Countries : Who should "Crowd Out" Whom in Imperfect Markets?'. In M. Cimoli, G. Dosi, and J. Stiglitz (eds), Industrial Policy and Development. Cambridge: Cambridge University Press.

Amsden, A. H., and W. W. Chu (2003) *Beyond Late Development: Taiwan's Upgrading Policies*. Cambridge, MA: MIT Press.

Amsden, A. H., and T. Hikino (1994) 'Project Execution Capability, Organizational Know-how and Conglomerate Corporate Growth in Late Industrialization', *Industrial and Corporate Change*, 3 (1): 111–47.

Bagchi, A. K. (1972) *Private Investment in India, 1900–1939*. Cambridge: Cambridge University Press.

Banerjee, A. K. (1963) *India's Balance of Payments: Estimates of Current and Capital Accounts from 1921–22 to 1938–39*. Bombay: Asia Publishing House.

Chandavarkar, R. (1994) *The Bombay Cotton Textile Industry, 1900–1940*. Cambridge: Cambridge University Press.

Das, N. (1962) *Industrial Enterprise in India*. Bombay: Orient Longmans.

Dosi, G. (1988) 'Sources, Procedures, and Microeconomic Effects of Innovation', *Journal of Economic Literature*, 26 (3): 1120–71.

Ferguson, N. (2003) *Empire: The Rise and Demise of the British World Order and the Lessons for Global Power*. New York: Basic Books.

Feuerwerker, A. (1967) 'Industrial Enterprise in Twentieth-Century China: The Chee Hsin Cement Co'. In A. Feuerwerker, R. Murphey, and M. C. Wright (eds), *Approaches to Modern Chinese History*. Berkeley and Los Angeles, CA: University of California Press.

Fraser, L. (1919) *Iron and Steel in India*. Bombay: Times Press.

Hao, Y.-P. (1986) *The Commercial Revolution in Nineteenth-Century China: The Rise of Sino-Western Mercantile Capitalism*. Berkeley, CA: University of California Press.

Khanna, T., and Y. Yafeh (2007) 'Business Groups in Emerging Markets: Paragons or Parasites?', *Journal of Economic Literature*, 45: 331–72.

Lawrence, R. Z. (1993) Japan's Low Levels of Inward Investment: The Role of Inhibitions on Acquisitions. In K. Froot (ed.), *Foreign Direct Investment*. Chicago, IL: University of Chicago Press for the National Bureau of Economic Research.

Maddison, A. (1971) *Class Structure and Economic Growth: India and Pakistan since the Moghuls*. New York: W. W. Norton.

Morris, M. D. (1983) 'The Growth of Large-Scale Industry to 1947'. In D. Kumar and M. Desai (eds), *The Cambridge Economic History of India*, Vol. 2. Cambridge: Cambridge University Press.

Mourshed, M. (1999) 'Technology Transfer Dynamics: Lessons from the Egyptian and Indian Pharmaceutical Industries', PhD Dissertation. Cambridge, MA: MIT.

Riskin, C. (1975) 'Surplus and Stagnation in Modern China'. In D. H. Perkins (ed.), *China's Modern Economy in Historical Perspective*. Stanford, CA: Stanford University Press.

Rutnagur, S. M. (1927) *Bombay Industries: The Cotton Mills*. Bombay: The Indian Textile Journal Ltd.

4

Types of Entrepreneurship and Economic Growth

Erik Stam and André van Stel

4.1 Introduction

Entrepreneurship has long been considered a crucial mechanism of economic development (Schumpeter 1934; Landes 1998). However, empirical studies on the role of entrepreneurship in economic growth show mixed evidence (Stam 2008). This is not remarkable because there is much heterogeneity in both the kinds of entrepreneurship and the kinds of economic contexts in which economic growth takes place. Until now studies have not sufficiently accounted for this heterogeneity on the micro- and macro-level, which limits our insight into the contingent role of entrepreneurship in economic growth. Important questions in this respect are: 'How does the role of entrepreneurship differ between high-income, transition, and medium-income countries?', and 'What kinds of entrepreneurship are most crucial for economic growth?'. The objective of this chapter is to provide insights into the role of different types of entrepreneurship in economic growth, and on how this role differs in poor and rich economies.

In this chapter, we empirically investigate the effect of entrepreneurship on economic growth at the country-level. We use data from the Global Entrepreneurship Monitor (GEM), which provides comparative data on entrepreneurship from a wide range of countries. An important element of this chapter is that we compare the effects of entrepreneurial activity on economic growth in high-income countries, transition countries (China, Hungary, Poland, Russia, and Slovenia), and medium-income countries (Argentina, Brazil, Chile, India, Mexico, South Africa, and Thailand). This dataset also enables us to make a distinction between the effects of entrepreneurship in general and growth-oriented entrepreneurship in particular. We present empirical tests of the impact of entrepreneurial activity on GDP growth over a four-year period for a sample of 36 countries.

Our empirical analyses suggest that entrepreneurship does not have an effect on economic growth in medium-income countries, in contrast to transition and high-income countries where especially growth-oriented entrepreneurship seems to contribute strongly to macroeconomic growth.

4.2 Entrepreneurship and economic development

Development is a broad concept entailing the raising of human capabilities (Sen 1999). One of the central challenges in improving economic development is to increase the standards of living for individuals and growth of the economy as a whole. Even though economic growth in itself is a rather narrow target, it is probably one of the most important targets for development policies. It is also one of the measures that is most easy to access for analysts, and probably the best measure to make cross-national (Barro 1991; Sala-i-Martin 1997) and historical (Maddison 2001) analyses of the development of economies. Traditionally the economic output of a country is seen as a function of capital and labour inputs, combined with technical change (Solow 1957). Of course, conflicts and wars might interrupt this function (Sala-i-Martin 1997), but these are 'just' contingencies. The standard production function used shows that economic output (Y) is a function of the sum of labour and capital inputs, and the level of technological knowledge (i.e. productivity). This means that economic growth—the growth of economic output—is a function of the growth of labour and capital inputs and technological progress. In traditional models of economic growth investment in capital, labour, and technology is sufficient to realize economic growth. New models of economic growth see these investments as a necessary complement to entrepreneurship/innovation, but not as a sufficient explanation for economic growth in its own right (Nelson and Pack 1999). One could even argue that high rates of investment in human and physical capital are themselves stimulated by effective innovation, and cannot be maintained in the absence of innovation. Recent studies emphasize entrepreneurship as a driver of economic development and some authors include entrepreneurship as a fourth production factor in the macroeconomic production function (Audretsch and Keilbach 2004).

Entrepreneurship is the factor that creates wealth by combining existing production factors in new ways. Entrepreneurs experiment with new combinations of which the outcomes are uncertain, but in order to make progress, many new variations have to be tried in order to find out which ones will improve (economic) life (Rosenberg and Birdzell 1986). Other authors have argued that entrepreneurship will only unlock economic development if a proper institutional setting is in place (Baumol 1990; Boettke and Coyne 2003; Powell 2008). This institutional setting comprises informal as well as formal institutions (North 1990). An essential formal institution for welfare enhancing entrepreneurship is property rights. Insecure property rights have been an important constraint on the investments by entrepreneurs in transition countries, even more so than capital market constraints (Johnson, McMillan, and Woodruff 2000). A specific example regarding property rights is the fact that until 1988 private firms with more than seven workers were not even allowed to operate legally in China (Dorn 2008). One might say that the production factors capital, labour, technology, and entrepreneurship are the proximate causes of economic development, while institutions are a fundamental cause of economic development (Acemoglu, Johnson, and Robinson 2004).

Next to productivity growth and technological change in established sectors, the development process in less advanced countries is largely about structural change (Gries and Naudé 2010; Nelson and Pack 1999; Rodrik 2007). It is a process in which an economy finds out—self-discovers—what it can be good at producing, out of the many products that already exist. The role of entrepreneurs in developing countries does not equal innovation and R&D as commonly understood in advanced economies. Their role is to discover that a certain good, already well-established in world markets, can be produced at home at low cost (Hausmann and Rodrik 2003; Rodrik 2007).[1] Examples of this are the entrepreneurs that figured out that Bangladesh was good in the production of T-shirts, Colombia in cut flowers, India in software services, and Taiwan in bicycles and display technologies. Even if entrepreneurs cannot appropriate all these gains for themselves, their discoveries generate large social gains for their economies. Spurring entrepreneurs to invest in their home economy is said to be one of the most important aspects of stimulating growth in poor countries (Rodrik 2007). Investing refers here to innovation (e.g. employing new technology, producing new products, searching for new markets) and expanding capacity. These investments trigger the combination of capital investment and technological change.

In advanced capitalist economies, innovation and structural change take place through the combined efforts of small (independent inventors) and large innovative (organized R&D) firms, which complement each other in changing the economy (Nooteboom 1994; Baumol 2002). In developing countries the role of large firms is relatively small (Ghoshal, Hahn, and Moran 1999). In transition countries there are relatively many large organizations but these are largely in a process of restructuring and dismantling. This means that small firms will be the prime movers in the process of structural change in developing and transition economies.

We expect that the level of growth-oriented entrepreneurship in a country is a more relevant driver of economic growth than the mostly used indicators of entrepreneurship like self-employment and new firm formation. In contrast to rich countries, entrepreneurship in medium-income countries is mainly driven by necessity (Bosma et al. 2008).[2] Most entrepreneurs in these economies do not start a firm because they desire independence or because they want to increase their income as compared to being an employee, which are the dominant motives in rich countries. Most new businesses in medium-income countries are started out of necessity, in contrast to high-income countries, where entrepreneurship is most often opportunity-driven. This is reflected in the finding that in poor countries self-employed persons are less happy than employees, while the reverse is true in high-income countries (Blanchflower and Oswald 1998; Graham 2005). Entrepreneurs in medium-income countries most often start a business because they have no other way of earning a living. These entrepreneurs are not likely to be involved in a process of self-discovery; their actions are not likely to have an effect on the restructuring and diversification of the poor economies (Rodrik 2007).

4.3 Data and research methods

It is generally acknowledged that there are differences in the distribution of entrepreneurship across countries. Studies exploring differences in entrepreneurship across countries often focus on the incidence of new firm registration or self-employment, which may not be reliable indicators when applied to transition and developing countries with significant informal economies and fewer alternatives to self-employment. For these reasons we have used the Young Business (YB) indicator, defined as the percentage of adult population that is the owner/manager of a business that is less than 42 months old. Many studies have used the total entrepreneurial activity index, but that also includes the more speculative category of nascent entrepreneurs (individuals preparing a new business). In the current study we investigate whether the presence of growth-oriented entrepreneurs is a more important determinant of national economic growth than entrepreneurial activity in general. We will perform regression analyses with the YB *high-growth expectation* rate and the YB *medium-growth expectation* rate as independent variables and compare their impact on economic growth with the impact of the general YB index. The data and model used in this study are described below.

We use a sample of 36 countries participating in the GEM in 2002. Data on six basic variables are used in our model: YB rate, YB medium-growth, YB high-growth, growth of GDP, per capita income, and the growth competitiveness index (GCI).

YB index

YB is defined as the percentage of adult population that is owner/manager of a business that is less than 42 months old. The YB high-(medium) growth expectation rate is defined as the percentage of adult population that is owner/manager of a business that is less than 42 months old, *and expects to employ 20 (six) employees or more within five years* (YB6 and YB20). The YB medium-growth rate has some similarity to the entrepreneurship indicator used by Djankov et al. (2006), which includes owner-managers of a business with five or more employees. Data on the YB rate are taken from the GEM Adult Population Survey for 2002.

Growth of GDP (ΔGDP)

(Real) GDP growth rates are taken from the IMF World Economic Outlook database of the International Monetary Fund from September 2005. In equations (1) and (2) below variable ΔGDP_{it} refers to the period 2002–5 (average annual growth) while the lagged GDP growth variable ($\Delta GDP_{i,t-1}$) refers to the period 1998–2001.

Per capita income (GNIC)

Most studies on GDP growth include the initial level of income in their analysis and find it to be significant (the conditional convergence effect, cf. Abramovitz 1986). Gross national income per capita 2001 is expressed in (thousands of) PPP dollars. These data are taken from the 2002 World Development Indicators database of the World Bank.

Growth Competitiveness Index (GCI)

In order to cover some aspects of the state of technology and institutions in a country (see Section 4.2) we used the GCI for the year 2001 of the World Economic Forum (see McArthur and Sachs 2002). Given the low number of observations we are forced to use a combined index in our model. Even though there are huge problems in measuring technological capabilities and institutions (see Lall 2001), the composite GCI is probably the best combined index available that covers these two factors simultaneously.

We investigate whether (growth-oriented) entrepreneurship may be considered as a determinant of economic growth, alongside the well-known determinants technology, institutions, and the macroeconomic environment, which are captured by the GCI. As both entrepreneurship and the factors underlying the GCI are assumed to be structural characteristics of an economy, we do not want to explain short-term economic growth but rather growth in the medium-term. Therefore we choose average annual growth over a period of four years (2002–5) as the dependent variable in this study. Following van Stel, Carree, and Thurik (2005), we use (the log of) initial income-level of countries to correct for catch-up effects, and lagged growth of GDP to correct for reversed causality effects, as additional control variables.[3]

We allow for the possibility of different effects for high-income, transition, and medium-income countries. In addition we also test whether the effect of YB is different for transition countries.[4] YB rates may reflect different types of entrepreneurs in countries with different development levels, implying different impacts on growth. This is tested by defining separate YB variables for different groups of countries (high-income, transition, and medium-income countries). Our model is represented by equations (1) and (2). These equations are estimated separately by ordinary least squares. The expectation that growth-oriented YBs contribute more to national economic growth than YBs in general corresponds to b_2 (c_2) being larger than b_1 (c_1). In these equations sub-scripts t and t-1 loosely indicate that the independent variables are measured prior to the dependent variable. The exact years and periods for which the variables are measured can be found in the variable description above.

$$\Delta GDP_{it} = a + b_1 YB^{rich}{}_{i,t-1} + c_1 YB^{transition}{}_{i,t-1} - d_1 YB^{poor}{}_{i,t-1} + e\log(GNIC_{i,t-1}) + f\,GCI_{i,t-1} + g\,\Delta GDP_{i,t-1} + \varepsilon_{it} \tag{1}$$

$$\Delta GDP_{it} = a + b_2 YB\_high - growth^{rich}{}_{i,t-1} + c_2 YB\_high - growth^{transition}{}_{i,t-1} + d_2 YB\_high - growth^{poor}{}_{i,t-1} + e\log(GNIC_{i,t-1}) + f\,GCI_{i,t-1} + g\,\Delta GDP_{i,t-1} + \varepsilon_{it} \tag{2}$$

To illustrate the data at hand, Table 4.1 provides the YB rates and the YB medium- and high-growth rates in 2002 as well as the average annual growth rates of GDP over the period 2002–5.

From Table 4.1 and Figures 4.1 and 4.2 it can be seen that the ranking of countries in terms of YB or YB high-growth may be quite different. For instance, while China ranks fifth in terms of YB, it ranks first in terms of

Table 4.1 Young business rates (2002) and GDP growth rates for 36 countries

	YB rate	YB medium-growth rate (6+)	YB high-growth rate (20+)	Average GDP growth rate 2002–5 (%)
United States	4.57	2.12	1.24	3.00
Russia	1.54	1.23	1.05	6.18
South Africa	2.00	0.88	0.58	3.60
Netherlands	2.09	0.90	0.63	0.60
Belgium	1.08	0.29	0.25	1.53
France	0.86	0.29	0.22	1.43
Spain	2.54	1.06	0.41	2.98
Hungary	3.62	1.29	0.98	3.50
Italy	2.35	1.14	0.84	0.48
Switzerland	3.26	1.28	0.38	0.60
United Kingdom	3.05	1.25	0.74	2.40
Denmark	3.12	1.43	0.58	1.45
Sweden	2.51	0.82	0.47	2.43
Norway	4.40	1.29	0.75	1.88
Poland	0.77	0.49	0.49	3.40
Germany	2.07	1.12	0.83	0.58
Mexico	3.22	0.81	0.32	2.40
Argentina	6.20	1.70	1.46	3.60
Brazil	8.46	3.17	2.34	2.65
Chile	5.49	3.83	2.23	4.48
Australia	5.22	2.10	1.25	3.18
New Zealand	6.06	2.50	1.50	3.85
Singapore	2.03	1.23	0.53	4.23
Thailand	8.40	2.52	1.37	5.45
Japan	1.04	0.52	0.26	1.45
Korea	9.29	3.95	2.43	4.63
China	7.41	2.83	2.57	9.08
India	7.45	2.68	2.15	6.63
Canada	3.58	1.51	0.91	2.73
Ireland	4.20	1.90	0.92	5.00
Iceland	6.23	2.79	1.98	3.28
Finland	2.06	0.71	0.43	2.50
Slovenia	1.53	0.71	0.41	3.58
Hong Kong	1.40	0.56	0.14	4.88
Taiwan	3.08	1.72	1.15	4.08
Israel	3.88	2.81	1.94	2.28
Mean	*3.78*	*1.60*	*1.02*	*3.22*

Sources: GEM and IMF.

high-growth YB. In contrast, Thailand ranks third in terms of YB, but only tenth in terms of high-growth YB.

When we regress the rate of GDP growth on the YB rate and the YB20 rate, the YB20 rate reveals to have a stronger correlation with GDP growth (see Figures 4.3 and 4.4).

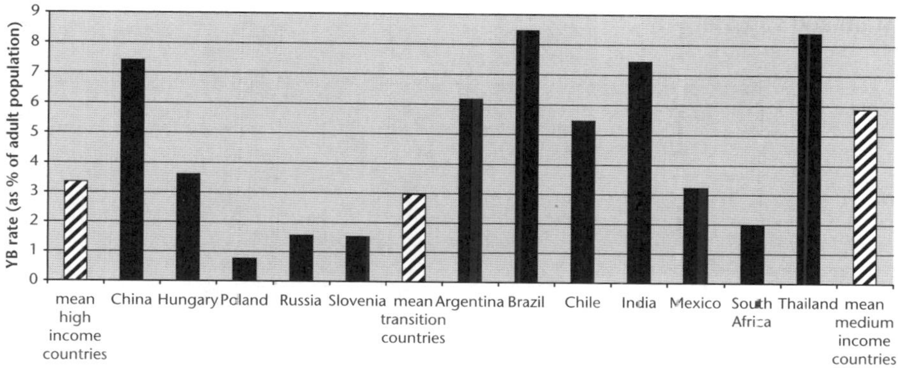

Figure 4.1 Young business rates

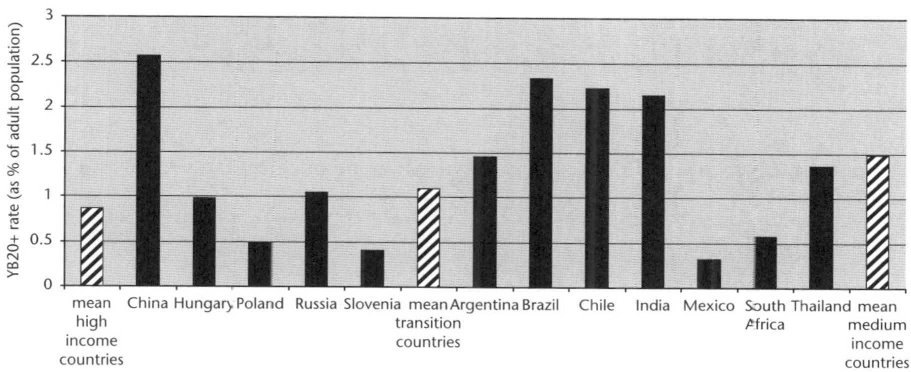

Figure 4.2 Young business 20 rates

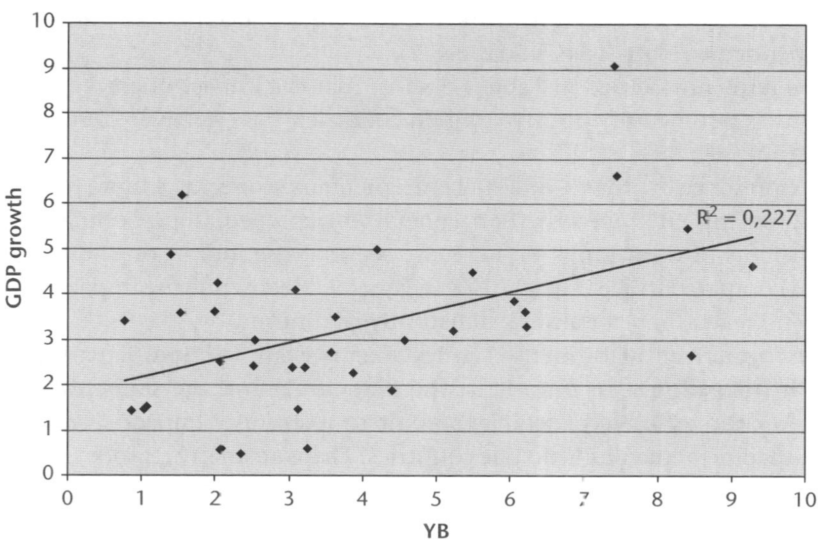

Figure 4.3 Correlation of young business rates and GDP growth rates

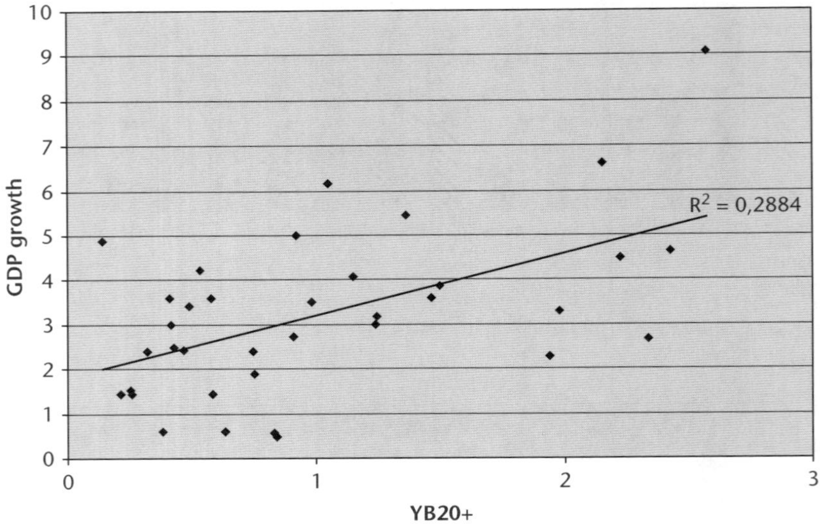

Figure 4.4 Correlation of high growth-oriented young business rates (20+) and GDP growth rates

4.4 Entrepreneurship and national economic growth

4.4.1 *Regression analyses*

The results of our empirical exercises are in Table 4.2. Model I presents the regression results of the impact of the general YB index (see equation (1)), while Models II and III show the results using the YB6 and YB20 rates as main independent variables (see equation (2)).

The results presented in Table 4.2 show that the impact of entrepreneurial activity is significantly positive for rich countries, but effectively zero for poor countries.

The presence of growth-oriented entrepreneurs seems to be more important for achieving GDP growth than general entrepreneurship. Comparing the coefficients of the various YB rates, we see that the impact of YB6 is greater when compared to the impact of YB in general. Meanwhile the impact of YB20 is even greater, but not always statistically significant.

Having more growth-oriented entrepreneurs seems to be particularly important in transition countries. Both the magnitude and the statistical significance of the estimated coefficient point to a stronger impact compared to high-income or medium-income countries. There are many reasons that could

Table 4.2 Regression models average annual growth of GDP over the period 2002–5 (N=36)

	Dependent variable average annual growth of GDP over the period 2002–5		
	Model I YB	Model II YB6	Model III YB20
Constant	21.3***	19.5***	18.9**
	(2.8)	(3.7)	(2.7)
Entrepreneurship in high-income countries†	0.20**	0.46**	0.48
	(2.6)	(2.4)	(1.3)
Entrepreneurship in transition countries	0.36**	1.24***	1.29**
	(2.1)	(3.2)	(2.5)
Entrepreneurship in medium-income countries	0.053	0.24	0.29
	(0.3)	(0.8)	(0.5)
Log (GNIC)	−2.3**	−2.2***	−2.2**
	(2.5)	(2.9)	(2.4)
GCI	0.59	0.80	0.86
	(0.7)	(1.1)	(1.1)
Lagged GDP growth	0.22	0.18	0.21
	(1.1)	(0.9)	(1.0)
R^2	0.672	0.693	0.676
Jarque–Bera statistic [p-value]	[0.259]	[0.278]	[0.427]

Notes: Absolute heteroskedasticity consistent t-values are between brackets.
* Significant at a 0.10 level;
** 0.05 level;
*** 0.01 level.
† interaction high-income country dummy with either YB, YB6, or YB20.

explain the importance of growth-oriented entrepreneurs in transition countries (Smallbone and Welter 2006). First, there are many entrepreneurial opportunities in formerly state-dominated sectors. Second, many highly qualified individuals lost their jobs at state-financed organizations (e.g. universities, enterprises, government services). Third, there are many highly qualified (potential) entrepreneurs in these countries (especially in Eastern European countries), who do not face the opportunity costs of working for large public or private organizations. Fourth, those highly qualified (potential) entrepreneurs are also well connected to the power networks that were, and to a large extent still are, important in the political and economic arena of these countries, which takes away some barriers for high-growth firms in these countries. Summarizing, it may be argued that in transition economies high-growth opportunities are more widely available and hence, a higher number of growth-oriented entrepreneurs willing to act on these opportunities may be particularly fruitful for achieving growth in these countries. However, we should be aware of the large diversity in the group of transition countries, which comprises countries like Russia and China, as well as Hungary and Slovenia.

Our regression results should be interpreted with care as the analysis is based on a limited number of observations (36 countries). As a test of robustness we estimated the models leaving out one country at a time, i.e. we computed 36 auxiliary regressions, where each regression uses 35 observations (each time leaving one of the 36 countries out). Although t-values sometimes dropped a little, coefficients and t-values were generally in line with those reported in Table 4.2. The country that matters the most for the results obtained in Table 4.2 is China. This is not surprising as China combines high YB/YB6/YB20 rates with high GDP growth rates (see Table 4.1). When leaving this country out of the sample, the coefficient (t-value) for the transition countries is 0.32 (0.5) for the YB rate, 1.47 (1.2) for the YB6 rate, and 1.72 (1.1) for the YB20 rate. The low t-values are in part due to the low number of observations. Note, however, that the coefficients are very similar to the full sample estimates reported in Table 4.2. Furthermore, the Jarque–Bera test on the normality of disturbances is passed for all models reported in Table 4.2, indicating that it is not necessary to remove individual country observations. Therefore we feel that our results are quite robust to the potential influence of outliers. Nevertheless, given the low number of observations, the results should only be seen as a first illustration of how the impact of different types of entrepreneurship may differ between groups of countries with different levels of development.

4.4.2 *Medium-income countries*

Within the groups of transition and developing economies there are substantial differences in entrepreneurship rates. Chile stands out because of a particularly high rate of growth-oriented entrepreneurship, while Mexico has a particularly low rate of growth-oriented entrepreneurship. In contrast to high-income countries, entrepreneurship in medium-income countries is mainly driven by necessity: self-employment is often the only occupational choice given a paucity of other sources of employment (necessity-based entrepreneurship; see Acs and Amoros 2008; Bosma et al. 2008). The actions of most of the entrepreneurs in medium-income countries are not likely to have an effect on the restructuring and diversification of the poor economies. This would be the whole story if the rates of growth-oriented entrepreneurship would also be marginal in these economies. This is only the case for Mexico. Next to Chile—where opportunity-driven entrepreneurship is dominant—Brazil, India, and Argentina perform quite well with respect to growth-oriented entrepreneurship. This means that there still is a substantial group of entrepreneurs in medium-income countries that might get involved in a process of self-discovery. The problem in practice is that in contrast to rich and transition economies, growth-oriented entrepreneurship is less likely to be realized in developing economies, due to constraints on the provision of capital and (skilled) labour.

An additional constraint in medium-income countries is that there are relatively few (foreign) large companies, which could act as a training ground for prospective growth-oriented entrepreneurs, and could open up distribution channels for new fledgling enterprises (Knorringa 1996). This is also reflected in the finding of Bosma, Stam, and Wennekers (2010) that the incidence of intrapreneurship (i.e. employees developing new business activities for their employer) is much lower in medium-income countries than in high-income countries. In addition, one should make a distinction between large firms with productive (manufacturing) and resource extractive (mining, oil) activities here, as the former will be more useful for the development of entrepreneurship than the latter.

4.4.3 *Transition countries*

New firms in transition countries not only displace obsolete incumbents but also fill in new markets, which were either non-existent or poorly populated in the past. Our study suggests that in transition countries, growth-oriented entrepreneurs make an important contribution to economic growth. They create new jobs with relatively high incomes which the small incumbent population of private firms cannot provide. This entrepreneurial growth process is facilitated by the relatively high levels of human capital in combination with relatively low opportunity costs of self-employment of the adult population. The high degree of environmental dynamism in these countries—which is likely to positively affect the level of growth expectations and realizations of entrepreneurs in these countries—requires ambitious and well-connected entrepreneurs in order to translate these abundant opportunities in economic growth.

There are considerable differences within the group of transition countries. Hsu (2005) shows that the role of these connections differs considerably between China and Russia: in China it was a tool which could be used to build enough trust to allow business transactions to succeed ('capitalism without contracts'). In contrast, in Russia these connections devolved into corruption, and faded in importance for ordinary citizens. Without a way to build trust or extend networks, Russians retreated into defensive involution, and engaged in predatory behaviour against those outside their small circles of friends. Instead of capitalism without contracts, Russia suffered the depredations of 'capitalists without capitalism'.

There are also substantial differences in entrepreneurship rates within the groups of transition economies. China stands out because of particularly high rates of growth-oriented entrepreneurship (cf. Hsu 2005). Even though the YB rate is below the average of transition countries, the growth of self-employment has been enormous, not only in the richer coastal provinces, but also in rural areas (Mohapatra, Rozelle, and Goodhue 2007). Research by Djankov et al. (2006) also shows that entrepreneurs in China are more risk-taking and more committed to an entrepreneurial career than entrepreneurs in Russia. In addition, Russia has (and had: see Hsu 2005) a particularly low rate of entrepreneurship in general as well. The striking difference between entrepreneurship rates in China and Russia can be explained by their different paths from socialism to capitalism: gradualism and a shock therapy (see Burawoy 1996).

In China the gradual transformation started with a policy of decollectivization (decentralization of property relations) in the late 1970s and the promotion of small-scale industry, with a focus on promoting independent entrepreneurship. Experimentation with new economic arrangements, for example privatization of small state-owned enterprises, has led to a favourable accumulation of productive capabilities in China. In contrast, Russia underwent a shock therapy in which the old communist regime was liquidated, with a focus on rapid privatization of the state sector. However, the Russian state failed to organize a market economy, which led to a coordination and entrepreneurial vacuum into which have stepped conglomerates, banks and mafia, siphoning off surplus from production to exchange (Burawoy 1996).

4.5 Discussion of policy implications

In this section we will briefly discuss the potential implications of our exploration of the relationship between types of entrepreneurship and economic growth for entrepreneurship policy and industrial/cluster policy in medium-income and transition countries.[5]

4.5.1 *Entrepreneurship policy*

Our empirical analyses suggest that entrepreneurship does not have an effect on economic growth in medium-income countries, in contrast to transition and high-income countries where both growth-oriented entrepreneurship and entrepreneurship in general seem to contribute strongly to macroeconomic growth. Does this mean that stimulating entrepreneurship in medium-income countries is bad policy? The least we can say is that stimulating entrepreneurship alone will be insufficient as it is likely to attract necessity entrepreneurs with low human capital levels who do not contribute to economic growth. The non-significant effect of entrepreneurship on economic growth in medium-income countries might point at a shortage of large firms in these countries. By exploitation of economies of scale and scope and by adopting and diffusing technology developed elsewhere, large firms are important in transforming a developing economy into a developed economy (van Stel, Carree, and Thurik 2005). In these economies local workers are more productive working as wage employees than as entrepreneurs. Nevertheless stimulating growth-oriented entrepreneurship might be an additional element of transforming a developing economy into a developed one. Attracting investments by large (possibly foreign) firms, stimulating growth-oriented entrepreneurship, investing in labour and capital, and improving the institutional framework may be the recipe for growth here. On the one hand this is old news, in that it provides a plea for the traditional role of governments to invest in education and physical infrastructure, and to build and maintain a set of institutions that enable the development of the private sector (cf. Rosenberg and Birdzell 1986). On the other hand, the addition of growth-oriented entrepreneurship in development policy for medium-income and transition countries is a new element. One must be careful to target the right group of entrepreneurs though, i.e. governments should avoid that resources made available through government stimulation programmes are absorbed by necessity entrepreneurs with low human capital levels.

4.5.2 *Industrial/cluster policy*

The focus of this chapter has been on the country-level, which disregards the sub-national level of analysis, and what is of particular relevance here, the level of regional clusters (regional concentrations of particular industries). These regional clusters have proved to be important drivers of economic development in, for example, Taiwan, India, and Brazil. These clusters are both driven by and drive growth-oriented entrepreneurship. Growth-oriented entrepreneurs that start to invest in a particular industry are needed in order to reach a critical mass that is needed to reach certain agglomeration economies (Braunerhjelm and Feldman 2006). If the build-up of capacity to this level of critical mass is not reached due to the lack of complementary investments, there might be a role to play for governments to overcome coordination failure, for example by providing investment guarantees for entrepreneurs (see Rodrik 2007). Such industrial policy is not about 'picking winners' or comprehensive planning, but encouraging experiments with new types of economic activity (Rodrik 2007). Since it is impossible to judge winners and losers in advance, competent and growth-oriented entrepreneurs should be encouraged to try, success should be rewarded, and failure should not be coddled (Nelson and Pack 1999).

These clusters do not have to be close to the technology frontier (as in advanced capitalist economies). The real policy implications arise from thinking carefully about the particular sources of advantage for a nascent cluster and why that source might yield short-term complements with the potential to become long-term substitutes (Bresnahan, Gambardella, and Saxenian 2001). Cooperation of clusters in developing countries with existing richer economies is not 'colonialist'. Take for example the linkages with the US. India and Taiwan are linked to the US (especially Silicon Valley) via outsourcing of software services and manufacturing (due to low labour costs), but also by a returning group of expatriates who have worked there, and who see the benefits of long distance collaboration (Saxenian 1999).

There is a flow of people—the so-called Argonauts (Saxenian 2006)—and ideas back and forth between rich and emerging economies. Migrant workers tend to be among the most entrepreneurial in society. Governments of developing countries should not only look at these expatriate workers as a source of remittances. Given their entrepreneurialism, skills, and exposure to business in the developed world, as well as the desire of many of them to return home, they may be very important as a source of self-discovery in their country of origin (Rodrik 2007). In addition to developing the private sector, these return migrants may provide the new elite needed for building up a civil society. Only a fraction of the money spent on attracting FDI would be needed to target nationals abroad. This would attract more knowledgeable human capital and durable investments than most FDI will do. Once critical mass is reached within a regional cluster, it is likely to generate or attract growth-oriented entrepreneurs (e.g. Argonauts), who in turn stimulate further macroeconomic growth.

4.5.3 *Limitations and further research*

The regression analyses in this chapter are of limited value: they have not only simplified the range[6] and (linear) effects of determinants for economic development, they have also dumbed down economic development to economic growth over a short-term (four-year) period. We know that sustaining growth is more difficult (and caused by different factors) than igniting it (Rodrik 2007). This also connects to one of the other shortcomings: sustaining growth probably requires much more extensive institutional reform than can be properly taken into account in linear regression analyses. Next to better measures of institutions, future research should take into account samples with low-income countries and a larger number of medium-income and transition countries, and multiple years in order to achieve more robust empirical analyses. In addition, our data did not allow for testing the multiplicative effect of entrepreneurship, so we only analysed the additive effect. A larger number of cases would enable the inclusion of the more traditional indicators of capital and labour in the analyses, and allow for testing the multiplicative effect.

Notes

1. In fragile and failed states, entrepreneurship has another role in building up the private sector. From another starting point, the same counts for formerly communist countries.
2. The only exceptions are Chile and Uruguay.
3. When the growth expectations for the national economy are good, more entrepreneurs may expect to expand their business in the years to come. Hence, there may also be a (reversed) effect of economic growth on (high expectation) entrepreneurship. To limit the potential impact of reversed causality we include lagged GDP growth as an additional explanatory variable. As this variable may influence both (high expectation) entrepreneurship and current GDP growth, omission might have led to a bias in the estimated effect of entrepreneurship on economic growth. We also measure YB rates in a year (2002) preceding the period over which the dependent variable is measured (2002–5). By including lagged independent variables as well as a lagged dependent variable on the right hand side, we basically measure the effects in a Granger-causality type of framework (Granger 1969). Still, the possibility of reversed effects cannot be ruled out completely.
4. The 36 countries in our sample are divided into three categories: high-income, medium-income, and transition. The 24 high-income countries are Australia, Belgium, Canada, Denmark, Finland, France, Germany, Hong Kong, Iceland, Ireland, Israel, Italy, Japan, Korea, the Netherlands, New Zealand, Norway, Singapore, Spain, Sweden, Switzerland, Taiwan, the UK, and the USA. Our seven medium-income nations are Argentina, Brazil, Chile, India, Mexico, South Africa, and Thailand. The five transition countries are China, Hungary, Poland, Russia, and Slovenia. With the exception of the high-income countries Hungary and Slovenia, the transitional countries can be classified as medium-income as well.
5. For a review of public policies for high-growth start ups in high-income countries, see Stam et al. (2007).
6. Even the usual suspects in models of economic growth, capital and labour growth, have been omitted here, and should be included in future more comprehensive analyses of the role of entrepreneurship in economic growth.

References

Abramovitz, M. (1986) 'Catching Up, Forging Ahead, and Falling Behind', *Journal of Economic History*, 46 (2): 385–406.

Acemoglu, D., S. Johnson, and J. Robinson (2004) 'Institutions as the Fundamental Cause of Long-run Growth', Working Paper 10481. Cambridge, MA: NBER.

Acs, Z., and J. E. Amoros (2008) 'Entrepreneurship and Competitiveness Dynamics in Latin America', *Small Business Economics*, 31 (3): 305–22.

Audretsch, D. B., and M. Keilbach (2004) 'Entrepreneurship Capital and Economic Performance', *Regional Studies*, 38: 949–59.

Barro, R. J. (1991) 'Economic Growth in a Cross Section of Countries', *Quarterly Journal of Economics*, May: 407–43.

Baumol, W. J. (1990) 'Entrepreneurship: Productive, Unproductive, and Destructive', *Journal of Political Economy*, 98 (5): 893–921.

——(2002) *The Free-Market Innovation Machine: Analyzing the Growth Miracle of Capitalism*. Princeton, NJ: Princeton University Press.

Blanchflower, D. G., and A. J. Oswald (1998) 'What Makes an Entrepreneur?', *Journal of Labor Economics*, 16 (1): 26–60.

Boettke, P., and C. J. Coyne (2003) 'Entrepreneurship and Development: Cause or Consequence?', *Advances in Austrian Economics*, 6: 67–88.

Bosma, N., K. Jones, E. Autio, and J. Levie (2008) *Global Entrepreneurship Monitor. 2007 Executive Report*. Babson Park, MA: Babson College, and London: London Business School.

Bosma, N., E. Stam, and S. Wennekers (2010) 'Intrapreneurship—An International Study', EIM Scales Paper H201005. Zoetermeer: EIM Business and Policy Research.

Braunerhjelm, P., and M. Feldman (eds) (2006) *Cluster Genesis. Technology-Based Industrial Development*. Oxford: Oxford University Press.

Bresnahan, T., A. Gambardella, and A. Saxenian (2001) 'Old Economy Inputs for New Economy Outputs: Cluster Formation in the New Silicon Valleys', *Industrial and Corporate Change*, 10 (4): 835–60.

Burawoy, M. (1996) 'The State and Economic Involution: Russia Through a China Lens', *World Development*, 24 (6): 1105–17.

Djankov, S., Y. Qian, G. Roland, and E. Zhuravskaya (2006) 'Entrepreneurship in China and Russia Compared', *Journal of the European Economic Association* 4 (2–3): 352–65.

Dorn, J. A. (2008) 'China's March Toward the Market'. In B. Powell (ed.), *Making Poor Nations Rich. Entrepreneurship and the Process of Economic Development*. Palo Alto, CA: Stanford University Press.

Ghoshal, S., M. Hahn, and P. Moran (1999) 'Management Competence, Firm Growth and Economic Progress', *Contributions to Political Economy*, 18: 121–50.

Graham, C. (2005) 'Insights on Development from the Economics of Happiness', *The World Bank Research Observer*, 20 (2): 201–31.

Granger, C. W. J. (1969) 'Investigating Causal Relations by Econometric Models and Cross-Spectral Methods', *Econometrica*, 37: 424–38.

Gries, T., and W. Naudé (2010) 'Entrepreneurship and Structural Economic Transformation', *Small Business Economics*, 34 (1): 13–29.

Hausmann, R., and D. Rodrik (2003) 'Economic Development as Self-discovery', *Journal of Development Economics*, 72: 603–33.

Hsu, C. (2005) 'Capitalism Without Contracts Versus Capitalists Without Capitalism: Comparing the Influence of Chinese Guanxi and Russian Blat on Marketization', *Communist and Post-Communist Studies*, 38: 309–27.

Johnson, S. H., J. McMillan, and C. M. Woodruff (2000) 'Entrepreneurs and the Ordering of Institutional Reform: Poland, Slovakia, Romania, Russia and Ukraine Compared', *The Economics of Transition*, 8 (1): 1–36.

Knorringa, P. (1996) *Economics of Collaboration. Indian Shoemakers Between Market and Hierarchy*. London: Sage.

Lall, S. (2001) 'Competitiveness Indices and Developing Countries: An Economic Evaluation of the Global Competitiveness Report', *World Development*, 29 (9): 1501–25.

Landes, D. S. (1998) *The Wealth and Poverty of Nations: Why Some Are So Rich and Some So Poor*. New York, NY: W. W. Norton.

Maddison, A. (2001) *The World Economy: A Millennial Perspective*. Paris: OECD.

McArthur, J. M, and J. D. Sachs (2002) *The Growth Competitiveness Index: Measuring Technological Advancement and the Stages of Development*. Geneva: World Economic Forum.

Mohapatra, S., S. D. Rozelle, and R. E. Goodhue (2007) 'Development or Distress? The Rise of Self-Employment in Rural China', *World Development*, 35 (1): 163–81.

Nelson, R. R., and H. Pack (1999) 'The Asian Miracle and Modern Growth Theory', *Economic Journal*, 109 (457): 416–36.

Nooteboom, B. (1994) 'Innovation and Diffusion in Small Firms: Theory and Evidence', *Small Business Economics*, 6: 327–47.

North, D. C. (1990) *Institutions, Institutional Change and Economic Performance*. Cambridge: Cambridge University Press.

Powell, B. (2008) *Making Poor Nations Rich. Entrepreneurship and the Process of Economic Development*. Palo Alto, CA: Stanford University Press.

Rodrik, D. (2007) *One Economics, Many Recipes: Globalization, Institutions, and Economic Growth*. Princeton, NJ: Princeton University Press.

Rosenberg, N., and L. Birdzell (1986) *How the West Grew Rich*. New York, NY: Basic Books.

Sala-i-Martin, X. X. (1997) 'I just ran two Million Regressions', *American Economic Review*, 87 (2): 178–83.

Saxenian, A. (1999) *Silicon Valley's New Immigrant Entrepreneurs*. San Francisco, CA: Public Policy Institute of California.

——(2006) *The New Argonauts. Regional Advantage in a Global Economy*. Cambridge, MA: Harvard University Press.

Schumpeter, J. (1934) *The Theory of Economic Development*. Cambridge, MA: Harvard University Press.

Sen, A. (1999) *Development as Freedom*. Oxford: Oxford University Press.

Smallbone, D., and F. Welter (2006) 'Conceptualising Entrepreneurship in a Transition Context', *International Journal of Entrepreneurship and Small Business*, 3 (2): 190–206.

Solow, R. M. (1957) 'Technical Change and the Aggregate Production Function', *The Review of Economics and Statistics*, 39 (3): 312–20.

Stam, E. (2008) 'Entrepreneurship and Innovation Policy'. In B. Nooteboom and E. Stam (eds), *Micro-Foundations for Innovation Policy*. Amsterdam: Amsterdam University Press and Chicago, IL: Chicago University Press.

Stam, E., K. Suddle, S. J. A. Hessels, and A. van Stel (2007) 'Los Emprendedores con Potencial de Crecimiento y el Desarrollo Económico. Politicas Públicas de Apoyo a los Emprendedores', *EKONOMIAZ, Revista Vasca de Economia*, 62 (2): 124–49.

Van Stel, A., M. Carree, and R. Thurik (2005) 'The Effect of Entrepreneurial Activity on National Economic Growth', *Small Business Economics*, 24: 311–21.

5

Can Small Firms Innovate? The Case of Clusters of Small Producers in Northern Vietnam

Jaap Voeten, Job de Haan, and Gerard de Groot

5.1 Introduction

Many economists, politicians, and economic actors consider innovation as the key to achieving competitiveness in today's globalized world. Although this viewpoint is generally accepted in economic circles, the question remains whether innovation is feasible and relevant for any firm in any economic reality. Is innovation within reach and a way to pursue for small, medium, as well as large firms, in developed and developing economies?

In current debates about globalization and competitiveness, innovation is often represented as providing opportunities and conditions for developing countries to participate in the world economy. Innovation is seen as a potential way in which low-income countries can strengthen their firms' competitive position within global value chains (Kaplinsky 2000; Gereffi, Humphrey, and Sturgeon 2005). Schmitz (1999) specifically refers to cases of clusters of small businesses in less developed countries that 'have broken into international markets'. Was this achievement the result of innovation?

Other authors argue that innovation is not feasible for small producers in the informal sector in low-income countries as a means to increase competitiveness. They assume that these producers will only play a limited role in formal economies, international markets, and globalization. Lewis's dual sector model of development (1954) included the 'trickle down' theory which assumes that economic growth and technology flow down from the wealthy at the top to the poor at the bottom. The appropriate technology approach (Schumacher 1973) urged Western development agencies to design simple technologies that would help poor small producers in low-income countries to escape from poverty. The indigenous knowledge approach takes the position that local knowledge and local markets should be tapped into. None of these approaches take local capacity for innovation into account. Rather they see such producers as being locked into patterns of traditional and indigenous ways of production. Even today, the Global Competitiveness Report 2006–7 cited in Caniëls and Romijn (2007) reflects a similar view: 'innovation is something that is only significantly undertaken once a country has reached a considerable level of economic advancement'. According to the report, innovation is not a particularly relevant, important or useful activity for the great majority of firms in low- and medium-income countries.

This chapter questions this assumption and analyses examples which suggest that innovation is a potential avenue for small producers in low-income countries. For instance in Northern Vietnam, several clusters of small producers engaged in traditional crafts have introduced new technologies, new products, and applied new business practices in recent years, expanding their sales on domestic and international markets. Conventional economic thought assumes that such traditional crafts will eventually disappear as a result of the modernization of these countries' economies, based on the belief that traditional production technologies are conventional and backward and not suited to global market conditions. However, the Vietnamese examples suggest otherwise and are the basis for further exploration of the extent to which these successes are the result of innovation and whether this has any broader implications.

If these examples from Vietnam are indeed innovation this would provide additional support for further research into the potential role of innovation in poor communities. To do so, it is informative to review the types of innovation, their features, similarities, organization, and how they emerge. Such insights can provide the basis for further theory building on the extent and significance of innovation in low-income countries.

However, before doing so, there is a methodological challenge that first needs to be addressed. How do we know whether something is actually an innovation? In economic theory today innovation is a very broad concept, largely defined in terms of Western economies. Many of those involved in studying innovation interpret its meaning in different ways. Moreover, the term innovation is not value free: innovation is 'hot' and virtually all social actors in Western economies today, whether they be firms, public services, or educational institutions, claim to be 'innovative'. Does contemporary economic theory, with its existing concepts and definitions, provide a suitable instrument for assessing innovation in clusters of small producers in developing countries?

5.2 Defining innovation: the theoretical framework

5.2.1 *Defining innovation*

Many authors have addressed the issue of defining innovation (Rogers 1998; Read 2000; Tether 2003; Szmytkowski 2005), although most acknowledge that defining innovation precisely is problematic. The difficulty is that innovation is an activity that is more complex than it at first appears: 'it is a serious mistake to treat an innovation as if it were a well-defined homogeneous thing that could be identified as entering the economy at a precise date' (Kline and Rosenberg 1986: 283). Despite much research into innovation in many fields, no single discipline has succeeded in uniting the fragmented thinking into one consistent umbrella theory, providing commonly agreed definitions and theoretical concepts.

The economic exploration of innovation started at the beginning of the twentieth century when neo-classical growth theories (Harrod 1939; Domar 1946) and production function models of Solow (1956) and Swan (1956) were unable to explain the actual dynamics of economic growth (Amable 1994). At that time, innovation was considered to be an exogenous variable, by nature a 'black box' (Rosenberg 1982). Technical change and innovation were outside the competence of classical economists and a domain for engineers and scientists (Freeman 1994).

Veblen (1904) was one of the first authors to challenge this position by stressing that the development of a new technology is not an exogenous force, but rather a set of material, economic, and social relationships shaped by businessmen, managers, and workers. Schumpeter (1934) incorporated and explicitly explained the term 'innovation' recognizing the direct link that exists between innovative activity and the dynamics of economic growth. He departed from the idea of an economic equilibrium theory and argued that innovator-entrepreneurs continuously changed the existing equilibrium by introducing newness (Brusoni, Cefis, and Orsenigo 2006). Schumpeter defined innovation as the introduction of new or improved products, production techniques, and organization structures as well the discovery of new markets, and the use of new input factors.

The 1970s and 1980s saw an increasing recognition of the difficulties of equilibrium theories, which assumed perfectly rational agents working within a static economic context (Dosi and Nelson 1994). As an alternative, Nelson and Winter (1977, 1982) proposed that economic growth through innovation could be understood as an evolutionary process, which is the endogenous outcome of an economic system (Romer 1994). Nelson and Winter (1977: 37) defined innovation broadly 'as a portmanteau to cover the wide range of variegated processes by which man's technologies evolve over time'. Dosi

(1988: 222) emphasized the process and learning element when defining innovation, which involved: 'the search for, and the discovery, experimentation, development, imitation and adoption of new products, new processes and new organizational set-ups'. Until today, most economic literature on innovation builds on the assumption that 'innovation is a process' (Lundvall 1992; Edquist 1997; Carayannis, Gonzalez, and Wetter 2003; Fagerberg, Mowery, and Nelson 2004; Szmytkowski 2005).

In the 1990s, Lundvall (1992), Freeman and Soete (1997), and Edquist (1997) argued that innovation should be analysed, not only in terms of a process of new and better techniques, but rather as a coevolutionary mechanism or system of technologies, organizations, and institutions. Freeman (1987) and Lundvall (1992) advanced the innovation system theory arguing that the innovation process takes place in a network of public and private sector institutions. Research in the field of economic geography further developed the learning dimension of the innovation systems approach applying it to describe learning-based regional production systems, also known as 'learning regions' (Rutten and Boekema 2007).

In the mid-1990s, attention to innovation in economic theory expanded enormously (Fagerberg and Verspagen 2006). Researchers from various economic backgrounds have increasingly discussed and analysed innovation in the context of globalization, since it is acknowledged that modern national economies are increasingly dominated by competitive global markets and growing dependency on international economic systems (Preissl and Solimene 2003). The notion that innovation ensures competitiveness through the creation of value has been important since Schumpeter. Porter (1990) again underlined the value creation and competitiveness aspects of innovation in his theory on new competitiveness. Firms create competitive advantage by perceiving or discovering new and better ways to compete and bringing them to market, which is, according to Porter (1990), the ultimate act of innovation. Value creation, profitability, and commercialization are key aspects of innovation in virtually all the definitions of innovation since Schumpeter (Krasner 1982; Edquist 1997; Rogers 1998; Walsh 2002; Fagerberg, Mowery, and Nelson 2004), and distinguish innovation from an invention. This implies that an innovation is by definition successful: innovation is the successful exploitation of ideas.

The theory and definitions discussed above repeatedly consider newness, value creation, and process as the key elements of innovation. Thus innovation can legitimately be summarized as the process of introducing something new that creates value. Today, innovation and entrepreneurship are at the forefront of academic debates in economics, business administration, and other related fields of study; they seem clearly interrelated and the role of the entrepreneur can only be understood if it is placed against the background of the theory of innovation. Hagedoorn (1996) referring to Schumpeter even states that the entrepreneur is the personification of innovation.

The three key elements of the innovation definition are still too broad to actually assess in practice whether something is an innovation. Further operationalization is necessary to assess whether something qualifies as 'new', whether this something new 'creates value', and whether the introduction of newness involves a 'process'.

5.2.2 *An instrument for assessing innovation*

There is substantial literature and quite a few approaches for assessing innovation. Most of the approaches measure either quantitative *output* of innovation (e.g. the number of patents obtained or the share of new products in total sales), or *input* into the innovation process, for example R&D expenditure or staff, or investment in innovation management (Freeman and Soete 2007). These approaches, however, cannot be used to measure the multidimensional definition of innovation, especially within clusters of informal small producers in developing countries, where it is generally difficult to obtain reliable quantitative business data. To address such a situation we need an assessment instrument that captures the multidimensional character of innovation and is context independent. Next, the chapter proposes a generic assessment instrument that uses a set of criteria and operationalization that are derived from the literature. This instrument also differentiates between the three key elements of innovation—newness, value creation, and process.

5.2.2.1 NEWNESS CRITERIA

Johannessen, Olsen, and Lumpkin (2001) observed that there is no agreement about the nature of newness. Yet, being a key element within virtually all definitions of innovation, some agreed criteria for newness are essential in identifying innovation.

Schumpeter (1934) defined six different types of innovative activity: new products, new services, new methods of production, opening new markets, new sources of supply, and new ways of organization. Johannessen, Olsen, and Lumpkin (2001) and Kaplinsky and Morris (2001) have reshaped the typology as: (i) process innovation—aiming at improving the efficiency of transforming inputs into outputs; (ii) product innovation leading to better quality, lower price and/or more differentiated products; (iii) business concept of practice innovation—new ways of doing business and attracting new clients. Kaplinsky and Morris (2001) include a further two categories taking over the functions of other actors in the value chain or switching to other chains altogether, (iv) functional innovations—assuming responsibility for new activities in the value chain; design, marketing, logistics, and (v) inter-chain innovations—moving to new and profitable chains.

Criterion 1.1	Threshold value
The new 'something' (newness) concerns one of the types of innovation agreed on in the literature (Schumpeter 1934; Kaplinsky and Morris 2001; Johannessen, Olsen, and Lumpkin 2001).	Newness can be classified either in terms of a new product, or process, or concept/practice, or function, or opening up a new market, or new sources of supply, or new ways of organization.

The next criterion concerns the application of the term newness. Chattopadhyay and Srivastava (2007) describe newness as what we have not encountered before. Newness exists where something is different from the past. There is a point in time that marks the arrival of newness. Johannessen, Olsen, and Lumpkin (2001) stress that newness is a relative, rather than an absolute, concept and here the question 'new to whom?' becomes important, since what is new to one firm could already exist somewhere else. Kotabe and Swan (1995) argue that innovation can be investigated in terms of both newness to the firm and newness to the market or world. The newness of something can only be assessed when the unit of analysis has been determined, for instance a firm or a cluster.

Criterion 1.2	Threshold value
The newness introduced represents a difference from the past within the specified unit of analysis (Kotabe and Swan 1995; Johannessen, Olsen, and Lumpkin 2001; Chattopadhyay and Srivastava 2007).	A point in time can be determined/identified that distinguishes between the times where the 'something new' did and did not exist in the unit of analysis.

The next question is how different or how new must something be to qualify as new? Most innovation studies acknowledge a distinction between incremental and radical innovations. The importance of incremental step-by-step innovation is often emphasized and much innovation is quite mundane, being incremental rather than radical (Freeman 1994). Much innovation depends more on an aggregation of small insights and advances through 'learning-by-doing' rather than on major technological inventions (Carayannis, Gonzalez, and Wetter 2003).

Since new is relative to the unit of analysis, setting an absolute scale of newness or a framework of reference is not possible. This subjectivity also implies that the newness should have a particular meaning to the people concerned. According to Porter (1990) innovation is the result of an unusual effort and doing something exceptional. People involved in innovating, whether producers or users, experience and acknowledge that the newness is a breakthrough with significance followed by 'adapters'.

Criterion 1.3	Threshold value
The producers and users perceive and acknowledge the newness as a breakthrough; a major achievement or success that permits further progress (Porter 1990; Freeman 1994).	It can be demonstrated that a few actors started to introduce the newness, later followed by others (early innovators → adopters) on a larger scale.

5.2.2.2 VALUE CREATION CRITERIA

The second element of the definition of innovation concerns value creation. According to Porter (1990), innovation generates value when a firm provides comparable value to buyers but performs its activities more efficiently through lower costs (cost advantage) or when a firm performs its activities in a unique way, thus creating greater buyer value and attracting a premium price (differentiation advantage). In other words, the newness can either lead to lower input costs or higher sales revenues.

Criterion 2.1	Threshold value
More value is added by the firm either through lower input costs or higher sales revenues (Porter 1985).	A causal explanation can be attributed between the introduction of the newness and lower input costs or higher sales revenues.

In addition to value creation within the firm, the literature on innovation also considers the impact of innovation on the firm's competitive advantage to be critical. Porter (1990) stresses the links between value creation and competitive advantage. Through innovations, firms can stay one step ahead of the competition. So another criterion of value creation is whether a firm is advancing its competitive position in the existing market (whether local, national, or international) or is able to enter into new, more profitable markets.

Criterion 2.2	Threshold value
More value is generated by improving / advancing the competitive position of the unit of analysis in local, national, or international markets (Porter 1985, 1990).	Market expansion and entry into new markets can be demonstrated after the introduction of the newness.

5.2.2.3 INNOVATION PROCESS CRITERIA

Initially, innovation was viewed as a one-dimensional 'linear process' proceeding sequentially through relatively independent steps: from research to marketing. This view overlooked the importance of feedback loops between research, technological knowledge, and the market. Dosi (1988) suggests that the essential steps include the discovery, experimentation, development, imitation, and adoption of something new. Tether (2003) sees the innovation process as typically starting with the generation of a creative idea or an invention, which is then brought to life through a research/test phase and an implementation phase: making an investment an essential part of the process. In sum, innovation is an unstructured process that follows a general pattern with three component elements: (i) creativity, ideas, or invention as solutions for the operation of the business; (ii) developing and testing a pilot, a prototype, a trial; and (iii) application, investment, implementation, and commercialization.

Criterion 3 1	Threshold value
The introduction of newness is typically an unstructured process of three component elements (Nelson and Winter 1982; Kline and Rosenberg 1986; Dosi and Nelson 1994; Tether 2003).	Within the unit of analysis, three component elements of the process can be identified: (i) creativity and the search for ideas; (ii) development and testing; and (i i) application, implementation, investment, and commercialization

Many authors confirm that innovation is a learning process. Dosi (1988) observed that a significant amount of innovations and improvements originate from 'learning-by-doing' and 'learning-by-using'. Mytelka and Smith (2001) observe that innovation research today has reconceptualized the firm as a learning organization focused on knowledge and learning.

Learning in an innovation process implies that an original idea is further improved in a cycle of loops, feedbacks, and checks in the three-step process described in criterion 3.1. Learning can be compared with walking through another cycle of these three elements.

Criterion 3.2	Threshold value
The introduction of newness is typically a learning process within the unit of analysis (Dosi 1988; Mytelka and Smith 2001).	Feedback during the process can be demonstrated to improve or build upon the original idea, and instigates another cycle/round of the 3-step process described in criterion 3.1.

Looking more closely at how learning takes place, Lundvall (1992), Freeman (1995), and Edquist (1997) advanced the theory that the process of innovation is characterized by interactive learning within an innovation system. It provides the interaction between the actors necessary for effective innovation.

Criterion 3.3	Threshold value
The innovation process is characterized by interaction in the environment of the unit of analysis (Freeman 1987, 1995; Lundvall 1992; Edquist 1997).	A causal attribution can be made between the introduction of newness and interactions beyond the unit of analysis.

The instrument for assessing innovation proposed here, therefore, involves testing eight criteria for a selected unit of analysis. Only if all criteria are met can the presence of innovation as a process of introducing something new that creates value be confirmed.

5.3 Analysing Vietnamese examples of new business dynamics

In 1986, Vietnam initiated an economic reform campaign (*Doi Moi*) setting in motion a transition process from a centrally planned to a free market economy. At that time, Vietnam was listed among the poorest countries in the world with per capita GDP at US$203. Since *Doi Moi*, the Vietnamese economy has experienced considerable growth as presented in Table 5.1. GDP growth averaged 7.8 per cent in the period 1995–2008 and per capita GDP quadrupled since the reforms were launched. The economic structure of Vietnam has changed significantly, with agriculture declining in importance from 40.8 per cent of GDP in 1989 to 20.1 per cent in 2006. Industry has gained proportionally in importance, growing from a percentage share of 22.9 per cent of GDP in 1989 to 40.1 per cent in 2006. During this period, the

Table 5.1 Selected indicators of economic growth in Vietnam, 1995–2008

	1995	2000	2001	2002	2003	2004	2005	2006	2007	2008
GDP growth (per cent)*	9.5	6.8	6.9	7.1	7.3	7.8	8.4	8.2	8.5	7.3
GDP (billions US$)**	20.7	31.1	32.5	35.1	39.6	45.5	53.0	60.9	70.0	81.3
GDP per capita (US$)**	288	401	413	440	489	555	637	722	818	937
Population (persons millions)	71.9	77.6	78.6	79.7	80.8	82.0	83.2	84.4	85.5	86.7

Note: * constant prices, ** current prices.

Source: International Monetary Fund (IMF), World Economic Outlook Database, April 2008 (available at: <http://www.imf.org/external/pubs/ft/weo/2008/01/weodata/index.aspx>).

Table 5.2 Establishments and employment in Vietnam, 2002

Vietnam (2002)		SME		Subtotal	Large SE	Total
Establishments	Micro	Small	Medium			
Number of business establishments (x 1000)	2,660	46.7	11	2,718	2.5	2,720
Percentage of all establishments (%)	97.8	1.7	0.4	99.9	0.09	100
Employment						
Employment (1000)	4,375	887	1,221	6,483	1,909	8,392
Percentage of persons engaged (%)	52.1	10.5	14.5	77.3	22.7	100
Average size of establishments						
Persons engaged per establishment	1.6	19	112	2.4	773	3

Source: GSO Establishments Census 2002, classified as per tentative size groupings (available at <www.gso.gov.vn>).

contribution of the services sector remained virtually unchanged at 36–8 per cent of the GDP.

The economic importance of micro, and small and medium-sized enterprises (SMEs[1]) has been significant in Vietnam in terms of number of businesses, employment creation (Table 5.2), and contribution to GDP. Many are engaged in retail trade, manufacturing, hospitality, and transportation while a number of small producers are located in clusters, similar to the cases selected as subjects for this study.

Recognizing the importance of SMEs in the economic development, the Vietnam government at present pays special attention to promotion and support for the development of micro- and household-based crafts businesses in the country. The new law on business enterprises and firms is aimed at promoting all kinds of businesses and economic sectors, creating equal conditions for everyone in business activities. The SMEs and in particular non-farm household enterprises are acknowledged to be important for their potential to absorb a growing labour force, to slow down regional and rural–urban migration, to promote a more equitable distribution of income (Oostendorp 2009).

Despite the economic advance during the last 20 years of economic reform, the formal Vietnamese private sector, and SMEs in particular, is not yet sufficiently competitive in a more global context (Nguyen et al. 2008). There are Vietnamese innovations and industrial policies focusing on specific support institutions for enterprises in technologically advanced industries, similar to Western innovation policy approaches. However, at the policy forum in 2006[2] 'Innovation Policies and Institutions for the Knowledge Economy', Vietnam was positioned among the group of countries that are only in the early stages of introducing innovation programmes for technology adoption and technology upgrading.

5.3.1 *Research methodology*

Since 1997, one of the authors of this chapter has been involved in training and research projects for household and SME development in Northern Vietnam. Typically, economic activities in this area revolve around agriculture and related activities. Also several villages have specialized in traditional crafts and small industries such as wood, silk, ceramics, noodles and so forth. Small producers in such villages often operate with some degree of cooperation, matching Schmitz's definition (1999: 466) of a cluster: 'the geographical and sectoral concentration of enterprises'.

Surveys showed the existence of several dynamic clusters that involved new ways of production, new products, and new business practices all of which enabled small producers to expand their markets. A variety of sources including development NGOs, media tourist agencies, and state economic agencies have all published reports with similar findings. These reports triggered our first explorations in identifying examples of clusters of small producers in 2006. The cluster was selected as unit of analysis for the case studies as this best represents the production system in these villages, given the interactions and interdependence of individual firms and the different roles the firms play as early innovators and late adaptors.

The initial data collection began with listing the craft villages and clusters of small producers through scanning various secondary resources: project reports, newspaper articles, internet sites, official and quasi-official documents, and a variety of resource persons. From this initial list a set of relevant clusters was short-listed for further exploration, with first field visits being carried out to more closely examine newness through observation and interviews with small producers so as to get a 'feel' for the new business dynamics atmosphere. This initial screening process led to the following clusters being selected for analysis:

- *Bat Trang*: a traditional ceramics village in the Red River Delta in Northern Vietnam, 15 km east of Hanoi. The village has 1,020 micro and small household enterprises[3] producing ceramics. Recently many small producers in the cluster introduced a new process—a gas oven—for baking ceramics and they have since expanded their market due to improved quality and increased production volume.
- *Duong Lieu*: a cassava starch and noodle-producing village in the Red River Delta, 30 km southwest of Hanoi. In the past five years, 20 small producer households switched from producing cassava noodles to a new end

product: children's sweets made from cassava. They now sell to more profitable outlet channels, such as supermarkets in Vietnam.

- *Van Phuc*: a traditional silk craft village in Ha Tay province, 10 km west of Hanoi where a cluster of 785 small, home-based producers is engaged in silk weaving, tailoring, and sales. Over the past ten years, many of these small producers have established retail shops in the village's main street, offering a much broader range of products.
- In the *Quang Hoa* district of the Thanh Hoa province, 225 km southwest of Hanoi, a development NGO started a technology transfer project in 2006 and established pre-processing workshops for small bamboo producers. Instead of selling unprocessed bamboo culms, small producers now cut, split, and smooth bamboo into slats for floor parts supplied to intermediaries of IKEA for the European market.

In May 2007, a second round of fieldwork took place. In-depth data collection focused on the assessment criteria through visual observations of the households, the workshops, the products, the tools and machines, and in-depth interviews with small producers, value chain suppliers and buyers, and local administration. After having collected sufficient empirical material for assessing the criteria, the data were further processed into case descriptions organized according to newness, value creation, and process, as described below. The case studies provided the basis for interpreting data for each criterion in the matrix presented at the end of this chapter. Finally, in January 2008, a third fieldwork trip was held to verify the case descriptions.

5.3.2 *Bat Trang*

NEWNESS

The first case concerns small producers in Bat Trang who traditionally produced pottery and ceramics in charcoal-briquette kilns. Over the past five years, two-thirds of them have switched to a technologically advanced gas oven. Better control of baking temperatures combined with more intense heat resulted in the production of thinner and smoother ceramics with fewer defects. While the assortment used to be limited to standard pottery and home ceramics, the new technology also allowed a broad variety of contemporary and popular design, types, shapes, and colours to be now produced.

In addition, the small producers started to take an active role in direct sales to new groups of clients through opening retail shops. Small producers linked up with tourist operators in Hanoi to promote Bat Trang, and in a short time the village has become a tourist destination for buying ceramics. Both the small producers and the local authorities consider the introduction of gas ovens in Bat Trang as a success story.

VALUE CREATION

The new developments translated into higher sales revenues for the small ceramics producers. The higher quality resulted in higher prices and the market expanded for domestic consumption, and increasingly for export contracts for Europe, Japan, and the US. Small producers play an important role in export through sub-contracts with larger companies and occasional direct contracts through tourists, families overseas, and individuals who visited the village.

PROCESS

The introduction of gas ovens was initiated by one small producer, Mr Le Duc Trong, who purchased a gas oven from China in 1995. Small producers in Bat Trang initially observed this with interest and slowly started to switch to a gas oven as well. After initial trials and testing, the small producers succeeded in getting the ovens to operate shortly after their installation and now produce and sell a larger volume of higher quality ceramics.

The small producers started to try out a broader assortment of products, picking up ideas from customers who suggested different shapes, designs, and colours for the ceramics. Typically, a producer first develops a few test samples, or produces some extra items of a contracted order and tests their utility and marketability before expanding production. Small producers compare results with other producers and review new technical possibilities and constraints, which determine the eventual selection of the assortment. The small producers are very aware of the need to do better all the time, not only because of increasing competition within their village but also from other villages that try to copy Bat Trang's success.

The local People's Committee actively promotes Bat Trang as the ceramics village and supports this through exposure, facilitating cooperation on business contacts, and infrastructure. Overseas families and friends advise on their preferences for product design and on technical matters. Some small producers have family contacts with the Polytechnic University in Hanoi, which conducts research on the quality of glazing.

5.3.3 *Duong Lieu*

NEWNESS

The second case concerns the introduction of a new product in Duong Lieu where many household businesses produce noodles from cassava starch. In the last five years, some 20 households have switched to producing a new end product: children's sweets from cassava starch.

Producing the sweets is a relatively basic and straightforward process that involves heating and mixing the cassava starch with several other ingredients.

The wrapping and packaging of sweets requires a major investment in a state-of-the-art machine. The small producers put effort into developing their own house-style for the packaging design. Several candy producers registered their designs at the Department of Property Rights, preventing others from copying them. Due to the considerable investment costs involved in organizing a new workshop set up, sweet production has, so far, only been feasible for middle-income households.

VALUE CREATION

Candy production adds more value to the processing of cassava starch than noodle production. The sweets are sold at a 'good' price to agents in Hanoi who distribute them to new profitable markets within Vietnam, such as shops, mini-markets, and supermarkets. The sweets sell well, especially at some holiday times. They compete with imported sweets and provide the households with higher overall sales revenues than those from noodles.

PROCESS

The initial idea for producing candy from starch came from one better-off family in the village. Today, this family business enterprise has become a successful small factory, serving as a model for other small candy producers. The switch to candy production implied an important change in the way in which workshops are set up, requiring investments in new equipment and machinery, redesigning the production line, and hiring new staff. All these steps were taken by the households themselves, without any external assistance.

The 20 candy producers currently have similar production facilities. There is a great deal of informal exchanging of ideas and practices within the cluster despite the fact that small producers consider their neighbouring sweets producers as competitors. The small producers therefore are continuously pursuing new types and tastes and consulting with the buying agents in Hanoi about new trends in taste, colour, and shapes, as well as in wrapping and packaging.

5.3.4 *Van Phuc*

NEWNESS

The third case concerns the introduction of a new marketing function. Before the introduction of the free market economy in Vietnam, silk products in Van Phuc were sold to state-owned intermediaries. In the 1990s, Ms Nguyen Truc Hong became the first person to open a shop selling local silk in the village. Many have followed her example and today there are over 100 silk shops in Van Phuc.

The producers also have broadened their range of products. Originally, the production focused exclusively on traditional silk fabrics, garments, accessories, and garnitures made from silk that they produced and tailored themselves. Nowadays one sees much more stylish design in the shops with new shapes, colours, designs, and a range of new products including shawls, jackets, pyjamas, sleeping bags, and accessories (like ties, bags, purses, etc.). Many of these new products break with the tradition of exclusively using high quality silk. Products are often mixed with synthetic materials of a lower quality.

VALUE CREATION

Over ten years, the overall silk production in Van Phuc has tripled and sales to domestic and foreign tourists visiting the small shops account for 40 per cent of the sales. The lower input costs and quality of the synthetic materials have resulted in lower prices, which have attracted new client groups who accept the lower quality. This has led to an overall increase in sales volumes.

PROCESS

The process of opening shops in the village started at the time when privately owned shops just began to develop in Vietnam. After the initial success of Ms Hong's retail shop, other small producers and traders followed and started to set up their own shops on an experimental basis; trying out different set-ups, product displays, and ranges. By closely watching whether clients come, what they buy, at what price, and what their neighbours did, the shop owners gradually improved their shops into attractive well-organized places, packed with a broad assortment of silk products, with salespeople able to provide information on the products, in English, if necessary.

The interactions within the cluster are critical; small producers keep an eye on each other's new product designs. Moreover they have developed informal networks with technical/education/vocational centres and links with tourist agencies in Hanoi that provide suggestions and feedback. The local authorities actively promote Van Phuc as a silk village and have invested in new infrastructure. The small producers are part of a larger silk industry in Vietnam, which includes fashion houses, large production and export companies, and government agencies. Ideas about design and so on are also gleaned from magazines, media, and other means.

5.3.5 *Quan Hoa*

NEWNESS

The fourth case concerns the introduction of bamboo pre-processing technology for small producers. In 2005, the French NGO 'Groupe de Recherche et d'Echanges Technologiques' (GRET) initiated a development project called the Bamboo Supply Chain Development project to improve the position of producers in the Quan Hoa and Ba Thuoc districts (Thanh Hoa Province, Northern Vietnam) in the bamboo value chain. Previously, the pre-processing steps were carried out by two larger bamboo factories—The Bamboo Factory (TBF) and Tien Dong—which did the cutting, splitting, and smoothing of bamboo into slats for further processing into floor parts, boards, and furniture components for export through IKEA to the European market.

The GRET project facilitated the establishment of three new slat production workshops and organized small bamboo producers' groups to operate and manage the workshops. The TBF and Tien Dong did not consider the workshops as competitors for their survival, but were cooperative and assisted the workshops by leasing them equipment and providing technical advice and specifications for the bamboo slat processing. Not long after the workshops' establishment, several other private initiatives emerged and copied the project workshop model and also began to supply slats to the bamboo factories.

VALUE CREATION

The underlying idea of the GRET project was that pre-processing bamboo into slats would provide the small producers with higher sales revenues. Despite the fact that value is added, the direct sales revenues still proved to be low due to the low prices offered by the bamboo factories TBF and Tien Dong, which are the buyers and leading players in setting the price of the bamboo. Alternative market channels have not yet been established. GRET continues to look for further technological developments for alternative by-products such as charcoal and mushroom growing substrate from bamboo saw dust.

PROCESS

Starting with the project idea in 2004, a team from GRET conducted a survey to explore the opportunities for, and feasibility of, slat production for bamboo producers. Subsequently, GRET facilitated the set-up of the workshops by proposing the appropriate technology and serving as a bridge linking the producers with the bamboo factories buying the processed slats. Once the workshops were established, the bamboo producers and technicians from GRET jointly tested and implemented the technology. Apart from some minor adjustments, the slat production process and machines have not changed since the establishment of the workshops.

The matrix below presents the interpretation and summary of the case descriptions for each criterion of the operationalized definition of innovation (described in section 5.2).

Box 5.1 CASE STUDIES AND INNOVATION CRITERIA

Unit of analysis: clusters of small producers	Bat Trang	Duong Lieu	Van Phuc	Quan Hoa
Criterion 1.1				
The new 'something' (newness) concerns one of the types of innovation agreed on in the literature.	The gas oven is a new production process enabling the production of higher volumes of higher quality, with more variety in design.	The production of sweets instead of noodles from starch is a more profitable new product.	Direct retail sales to new client groups are taking over the marketing function from other players in the value chain.	The pre-processing of bamboo poles into slats is a new function applied by small producers.
Criterion satisfied	Yes	Yes	Yes	Yes
Criterion 1.2				
The newness introduced represents a difference from the past within the specified unit of analysis.	The first small producers purchased the gas ovens in 2001/2. Before that, ceramics in Bat Trang were only produced in charcoal kilns.	Five years ago a cluster of small producers started to produce the candy. One candy factory was established in the village 13 years ago.	The first shops were established in 1995. Before that time it was difficult to set up a private shop in Vietnam.	In 2005 GRET started to establish three slats workshops. Before then there was only one existing workshop producing chop sticks.
Criterion satisfied	Yes	Yes	Yes	Yes
Criterion 1.3				
The producers and users perceive and acknowledge the newness as a breakthrough; a major achievement or success that permits further progress.	Over the past six years the gas oven has been adopted by two-thirds of all small producers in Bat Trang.	Over five years, 20 households have switched to candy production and there is evidence of a growing trend in the village to switch to candy production.	Nearly every house on the main street has transformed into a retail shop since 1995. At present there are around 100 silk shops.	Several private initiatives have copied the workshop example and are now producing floor parts.
Criterion satisfied	Yes	Yes	Yes	Yes
Criterion 2.1				
More value is added by the firm either through	Higher sales revenues as a result of the	The production of sweets instead	Higher sales revenues as a result of higher	Higher sales revenues as a result of the

lower input costs or higher sales revenues.	increase in quality of the ceramics. Greater buyer value implying a higher price.	of noodles results in higher sales revenues.	sales volumes and lower input costs for mixed silk fabrics.	pre-processing of bamboo into strips.
Criterion satisfied	Yes	Yes	Yes	Yes
Criterion 2.2				
More value is generated by improving the competitive position of the unit of analysis in local, national, or international markets.	New customers such as foreign tourists, restaurants, and hotels. These occasionally enter into follow-up contracts with Japanese, European, and American visitors.	Although both noodles and sweets are sold on the domestic market, the sweets are sold into new and more profitable markets, such as supermarkets in Hanoi.	New and broader client groups—both domestic and foreign tourists—are coming to Van Phuc to buy silk and silk products.	The small producers did not enter new markets and their competitive position has not really changed. The workshops only can sell to 2 buyers as there is a high level of vertical integration in the chain.
Criterion satisfied	Yes	Yes	Yes	No
Criterion 3.1				
The introduction of the newness is typically a chaotic process of three component elements.	The idea of the gas oven came from the small producers themselves with one producer taking the initial step of purchasing one. The small producers experimented with the best way to operate the oven before producing on a larger scale and commercializing products.	The small producers themselves got the idea to switch to candy production and did the exploratory and preparatory work themselves. They tested whether they could successfully sell the candies, and started to explore ideas to improve sales / margins by using attractive wrappers.	The idea of establishing shops came from within the village. Gradually shops were set up, and improved. Shop owners continue to test new ideas to make their shops as attractive as possible, including having the workshop nearby so that tourists can visit.	The workshop owners themselves did not go through the three stages described in the literature. Ideas were imported from outside, which also supplied the machinery and production standards. The owners were only involved in the implementation phase.
Criterion satisfied	Yes	Yes	Yes	No

continued

Box 5.1 continued

Criterion 3.2

The introduction of newness is typically a learning process within the unit of analysis.	Small producers continue to seek to improve the quality of their ceramic products. They continuously generate ideas for better glazing, test these, and implement them if they prove successful.	Small producers test new textures, tastes, colours, and wrapping of the sweets. Every year buyers ask for new flavours and the producers respond to these demands.	The shop owners pursue new ideas and experiment themselves to make the shops more attractive and select the best range of products, which are constantly evolving.	The farmers did not further develop the strip processing machine technology and still use it the same way as when it was originally installed.
Criterion satisfied	Yes	Yes	Yes	No

Criterion 3.3

The innovation process is characterized by interaction in the environment of the unit of analysis.	There is interaction with buyers who suggest designs, colours, and the quality of the ceramic products. The authorities support ceramics production in Bat Trang and universities do research in glazing techniques.	Interaction with buyers, mostly in Hanoi, about the taste of the sweets.	There is interaction with tour operators, and with the clients who suggest products. The local authorities and national government are promoting Van Phuc as a silk village. There are exchanges with fashion schools.	There is interaction with the development NGO and the factories that buy the bamboo strips.
Criterion satisfied	Yes	Yes	Yes	Yes
All criteria satisfied?	Yes	Yes	Yes	No

5.4 Discussion and conclusions

5.4.1 *Discussion*

The primary empirical data presented in Box 5.1 allow us to give a positive answer to the question whether innovation occurs in clusters of small producers in Northern Vietnam: process innovation takes place in Bat Trang ceramics village, product innovation in Duong Lieu village, and functional innovation in Van Phuc silk village. The results confirm innovation at the 'cluster-level'—the unit of analysis. The conclusion that innovation does take place in these three traditional Vietnamese craft villages is perhaps surprising in the sense that innovation was neither planned for nor promoted, and no formal innovation system exists with a specific agenda for promoting innovation.

The fourth case study did not meet all the criteria. Contrary to the expectation of this technology transfer project for bamboo small producers, 'cluster-level' innovation did not take place. Several steps of the innovation process and the associated learning took place outside the cluster by the international development NGO. Another criterion not met by this case was improving its competitive position. After adding the new production steps the products were still sold to the same buyers in the value chain holding strong bargaining power. As a result the actual competitive position of the groups did not change and there was little value creation effect.

The small producers innovated by themselves, drawing on their own strengths and initiative via internal processes, interactions, and knowledge accumulation within the cluster. In this respect, this chapter demonstrates the local innovation capacity of small producers who supplement and combine local indigenous knowledge and technologies with global state-of-the-art technology. This contradicts the underlying assumptions of the trickle down theory, appropriate technology, and indigenous knowledge for low-income countries that do not adequately acknowledge local innovation capacity. The fourth case study is in line with these theories advocating external assistance and interventions to help small producers learn and advance. However, the external actor was eager and diligent to take over and own the learning in the innovation process. The question what influences eagerness to learn and discover is not yet understood within evolutionary economics, even though the discipline recognizes that learning is a critical element in the innovation process (Dosi and Nelson 1994). This also is relevant for these poor communities: For what reason do they have a particular drive to innovate? Is there an optimum or 'desirable' level of learning or an optimum amount of innovation?

The absence of direct external public or private 'innovation' support or interventions in the three successful examples is in line with endogenous growth theory (Romer 1994) which argues that economic growth comes from within a system. However, though the innovation process takes place entirely within the clusters, there was also much interaction with external environment: ideas, suggestions, and opportunities came from buyers, sellers, media, and industries. This implies a need for further understanding what role endogenous growth and innovation processes play in such contexts and the relevant contributions made by internal and external factors.

The absence of external innovation support actors can also be viewed from the innovations systems theory perspective. This study presents three cases of innovation that occur where there is no system of formal public and private organizations actively and deliberately promoting innovation. Moreover, the steps of the innovation process in these examples took place in the informally structured clusters, raising the question of whether the cluster itself constitutes an informal system of innovation. In the three confirmed innovation cases there are interactions with a larger system but this does not involve sharing or owning innovation process steps—as formulated in the innovation system definition—but merely exchanging incentives, ideas, and suggestions from clients, suppliers, competitors, etc. If the interactions do not involve sharing steps in the innovation process, then how precisely do these interactions fit within the innovation systems theory? The operationalization of the definition shows the necessity to distinguish between a 'shared innovation process' and 'interaction'.

How is this shared innovation process structured in a larger system? Edquist (1997) emphasized that institutions play an increasingly important role in innovation systems theory. As the number of actors involved increases, the innovation process becomes more complicated and more interactions occur. For both informal and formal innovation systems, questions about how these systems are organized emerge. How are the interactions and the cumulative knowledge generation of the system's actors structured? How is the created value shared within the system?

Regarding the sharing of value creation, Gereffi, Humphrey, and Sturgeon (2005) take the position that innovation can enable low-income countries to strengthen the competitiveness of their firms through participation in global value chains. The fourth case describes small producers taking over a bamboo pre-processing function from the leading actors in the chain, which could be labelled as outsourcing. It illustrates how the created value is appropriated by the lead actor. New technologies may be introduced to small producers in the value chain, but if improved competitiveness does not materialize then, according to the operationalized definition, this is not innovation.

Another issue for further discussion is the usefulness of the innovation assessment instrument itself as a research tool. Although the theoretical basis of the instrument comes from contemporary economic concepts on innovation, which were principally developed from studies rooted in the context of Western developed economies, the operationalized definition was able to differentiate between innovation and non-innovation in the context of a low-income country. This confirms the validity of the conclusion, that innovation did actually take place in three of the four cases.

There is scope for further refinement of the instrument's criteria and operationalization. Since these criteria have not been completely developed in the literature, it was not possible to make unambiguous choices for all of them. The operationalization of whether all steps in the innovation process take place within the cluster requires a detailed historical review. Different people involved could have different perceptions of the past, making it difficult to reconstruct the historical path. Assessing the empirical data against the operationalization for learning also proved to be challenging: at what point is the learning involved related to innovation?

The operationalization of the criteria illustrates the need to be explicit about the level at which innovation is assessed: the firm, the cluster, the value chain and so forth. At one level the research instrument might confirm a criterion, while at another level it might not. For instance, if the unit of analysis of the fourth case is altered to a broader level—incorporating the development NGO that introduced the newness, then the process and learning criteria would be confirmed. At the same time, other criteria may cease to apply when the unit of analysis is enlarged. For example, in the fourth case, the production process was new for the cluster, but for the broader level the newness criterion would not be confirmed.

5.4.2 *Summary and conclusions*

This chapter uses a newly developed research instrument to measure innovation in a developing country context. It demonstrates that innovation can and does occur in small producers' clusters in a developing country. Poorer producers take advantage of domestic economic growth and globalization. This is not in line with some positions in the development economics debate arguing that globalization mostly widens the gap between rich and poor. This is certainly an interesting avenue to get a deeper understanding of innovation and development processes in low-income countries, raising the following suggestions for a future research agenda.

The first issue for further research concerns the explanation of the emergence of innovation. In three of our cases small producers innovated on their own account, using their own strengths and initiative, while the case of the 'technology transfer project' did not demonstrate innovation. Comprehensive lists of innovation processes, factors, and drivers have been described for Western companies, but what about clusters of small producers in Vietnam? How did the innovation emerge in the confirmed cases? Do similar factors apply? Related questions include: what made it possible for small producers to innovate on their own strengths without the support of an innovation system, understood as necessary in Western economies? Was it because of endogenous or exogenous factors? Does this suggest the existence of some kind of informal innovation system? And, what determines the eagerness of small producers to learn and innovate?

A second issue is the contribution that innovation makes to poverty alleviation in a broader context. This chapter reviews three success stories of innovation, but does not analyse the effects on, for instance, neighbouring communities that did not introduce new things. Was the success of the three villages realized at the expense of other villages in the neighbourhood? How many failing villages will there be for every success story? Equally, within the cluster there can be a question of the distribution of the benefits, particularly given the heterogeneity within the clusters of small producers or within the value chain. Do the early innovators take a disproportionate advantage of the value created?

Finally, the operationalization of the definition of innovation helped us explore innovation among clusters of small producers in a developing country. Further research and a broader application of the research instrument could refine the operationalization and allow us to assess the scope for innovation among small producers on a larger scale, providing comparative data across sectors, geographic areas or businesses at various stages of development. When more such studies from developing countries become available, the question 'is it innovation?' can be addressed more systematically by drawing on a body of literature and empirical data that studies innovation in developing countries.

Notes

1. SMEs have been divided into three sub-groups: (i) microenterprises: engaging up to nine employees; (ii) small enterprises, engaging up to 49 employees; (iii) medium-size enterprises, engaging up to 299 employees.
2. Held at the Incheon Education and Science Research Institute, Seoul, Republic of Korea, 29 November–1 December 2006.
3. Micro and small entrepreneurs in Bat Trang typically have a home-based workshop, with between one and five (micro) or five and 20 (small) employees, often family members employed under informal contracts.

References

Amable, B. (1994) 'Endogenous Growth Theory, Convergence and Divergence'. In G. Silverberg and L. Soete (eds), *The Economics of Growth and Technical Change—Technologies, Nations, Agents*. Cheltenham: Edward Elgar.

Brusoni, S., E. Cefis, and L. Orsenigo (2006) 'Innovate or Die? A Critical Review of the Literature on Innovation and Performance', WP 179 Milan: Università Commerciale 'Luigi Bocconi', Centro di Ricerca sui Processi di Innovazione e Internazionalizzazione.

Caniëls, M. C. J., and H. Romijn (2007) 'Does Innovation Matter for LDCs? Discussion and New Agenda'. Paper presented at the Centre for Advanced Study (CAS) Workshop 'Innovation in Firms', Oslo, Norway, 30 October–1 November 2007.

Carayannis, E. G., E. Gonzalez, and J. Wetter (2003) 'The Nature and Dynamics of Discontinuous and Disruptive Innovations from a Learning and Knowledge Management Perspective'. In L. Shavinia (ed.), *The International Handbook of Innovation*. Kidlington: Elsevier.

Chattopadhyay, D., and B. N. Srivastava (2007) 'The Concept of Newness', *International Journal of Management Concepts and Philosophy*, 2 (3): 240–5.

Domar, R. (1946) 'Capital Expansion, Rate of Growth and Employment', *Econometrica*, April, 137–47.

Dosi, G. (1988) 'The Nature of the Innovation Process'. In G. Dosi, C. Freeman, R. Nelson, G. Silverberg, and L. Soete (eds), *Technical Change and Economic Theory*. London/New York: Pinter.

Dosi, G., and R. R. Nelson (1994) 'An Introduction to Evolutionary Theories in Economics', *Journal of Evolutionary Economics*, 4: 153–72.

Edquist, C. (1997) *Systems of Innovation: Technologies, Institutions and Organisations*. London: Pinter.

Fagerberg J., D. C. Mowery, and R. R. Nelson (eds) (2004) *The Oxford Handbook of Innovation*. Oxford: Oxford University Press.

Fagerberg, J., and B. Verspagen (2006) 'Innovation Studies——⟩An Emerging Discipline (or What)? A Study of the Global Network of Innovation Scholars'. Paper presented at the SPRU 40th Anniversary Conference, The Future of Science, Technology and Innovation Policy, Sussex, UK, 11–13 September 2006.

——. (1994) 'The Economics of Technical Change: A Critical Survey', *Cambridge Journal of Economics*, 18 (4): 463–514.

Freeman, C. (1987) *Technology Policy and Economic Performance: Lessons from Japan*. London: Pinter.

—— (1995) 'The National System of Innovation in Historical Perspective', *Cambridge Journal of Economics*, 19: 5–24.

Freeman, C., and L. Soete (1997) *The Economics of Industrial Innovation*, 3rd edn. Cambridge, MA: MIT Press.

Freeman, C., and L. Soete (2007) 'Developing Science, Technology and Innovation Indicators: What We Can Learn From The Past', Working Paper 2007-001. Maastricht: UNU-MERIT.

Gereffi, G., J. Humphrey, and T. Sturgeon (2005) 'The Governance of Global Value Chains', *Review of International Political Economy*, 12 (1): 78–104.

Hagedoorn, J. (1996) 'Innovation and Entrepreneurship: Schumpeter Revisited', *Industrial and Corporate Change*, 5 (3): 883–96.

Harrod, R. F. (1939) 'An Essay in Dynamic Theory', *Economic Journal*, March: 14–33.

Johannessen, J.-A., B. Olsen, and G. T. Lumpkin (2001) 'Innovation as Newness: What is New, How New, and New to Whom?', *European Journal of Innovation Management*, 4 (1): 20–31.

Kaplinsky, R. (2000) 'Spreading the Gains from Globalisation: What can be Learned from Value Chain Analysis?, *Journal of Development Studies*, 37 (2): 117–46.

Kaplinsky, R., and M. Morris (2001) *A Handbook for Value Chain Research*. Brighton: Institute of Development Studies.

Kline, S. J., and N. Rosenberg (1986) 'An Overview of Innovation'. In R. Landau and N. Rosenberg (eds), *The Positive Sum Strategy: Harnessing Technology for Economic Growth*. Washington, DC: National Academies Press.

Kotabe, M., and K. S. Swan (1995) 'The Role of Strategic Alliances in High Technology New Product Development', *Strategic Management Journal*, 16 (8): 621–36.

Krasner, O. J. (1982) 'The Role of Entrepreneurs in Innovation'. In C. A. Kent, D. L. Sexton, and K. H. Vesper (eds), *Encyclopaedia of Entrepreneurship*. Englewood Cliffs, NJ: Prentice Hall.

Lewis, W. A. (1954) 'Economic Development with Unlimited Supplies of Labor', *Journal of the Manchester School of Economics and Social Studies*, 20: 139–92.

Lundvall, B.-A. (ed.) (1992) *National Systems of Innovation: Towards a Theory of Innovation and Interactive Learning*. London: Pinter.

Mytelka, L., and K. Smith (2001) 'Innnovation Theory and Innovation Policy: Bridging the Gap'. Paper presented at the Nelson and Winter DRUID Conference, Aalborg, Denmark, 12–15 June.

Nelson, R. R., and S. G. Winter (1977) 'In Search of Useful Theory of Innovation', *Research Policy*, 6 (1): 36–76.

—— (1982) *An Evolutionary Theory of Economic Change*. Cambridge: Bellknap Press.

Nguyen, A. N., N. Q. Pham, C. D. Nguyen, and N. D. Nguyen (2008) 'Innovation and Exports in Vietnam's SME Sector', *The European Journal of Development Research*, 20 (2): 262–80.

North, D. C. (1990) *Institutions, Institutional Change and Economic Performance*. Cambridge: Cambridge University Press.

Oostendorp, M. H. (2009) 'The Changing Role of Non-farm Household Enterprises in Vietnam', *World Development*, 37 (3): 632–44.

Porter, M. (1985) *Competitive Advantage*. New York, NY: Free Press.

—— (1990) *The Competitive Advantage of Nations*. London: Macmillan.

Preissl, B., and L. Solimene (2003) *The Dynamics of Clusters and Innovation—Beyond Systems and Networks*. Heidelberg: Physica Verlag.

Read, A. (2000) 'Determinants of Successful Organisational Innovation: A Review of Current Research', *Journal of Management Practice*, 3 (1): 95–119.

Rogers, M. (1998) 'The Definition and Measurement of Innovation', Working Paper 10/98. Melbourne: Melbourne Institute of Applied Economic and Social Research.

Romer, P. M. (1994) 'The Origins of Endogenous Growth', *Journal of Economic Perspectives*, 8 (1): 3–22.

Rosenberg, N. (ed.) (1982) *Inside the Black Box: Technology and Economics*. Cambridge: Cambridge University Press.

Rutten, R., and F. Boekema (2007) *The Learning Region: Foundations, State of the Art, Future*. Cheltenham: Edward Elgar.

Schmitz, H. (1999) 'Collective Efficiency and Increasing Returns'. *Cambridge Journal of Economics*, 23: 465–83.

Schumacher, E. F. (1973) *Small Is Beautiful: Economics As If People Mattered*. New York, NY: Harper Collins.

Schumpeter, J. (1934) *The Theory of Economic Development*. Cambridge, MA: Harvard University Press.

Solow, R. M. (1956) 'A Contribution to the Theory of Economic Growth', *Quarterly Journal of Economics*, 70: 65–94.

Swan, T. W. (1956) 'Economic and Capital Accumulation', *Economic Record*, 32: 334–61.

Szmytkowski, D. (2005) *Innovation Definition Comparative Assessment*. Brussels: GNU Free Documentation License.

Tether, B. S. (2003) What is Innovation?—Approaches to Distinguishing New Products and Processes from Existing Products and Processes', Working Paper 12. Manchester: ESRC Centre for Research on Innovation and Competition (CRIC).

Veblen, T. (1904) *The Theory of the Business Enterprise*. London: Routledge/Thoemmes Press.

Walsh, V. (2002) Brief History of Economic Thought: Demand, Markets, and Selection Environments. Lecture. Manchester: Manchester School of Management.

6

Innovation and Microenterprise Growth in Ethiopia

Mulu Gebreeyesus

6.1 Introduction

In developing countries the informal sector which mainly constitutes microenterprises is the major source of employment and income for the urban population. According to ILO (2002) estimations, the share of informal employment (outside agriculture) accounts for nearly half or more of non-agricultural employment in all regions of the developing world and for about 72 per cent in sub-Saharan Africa (SSA). In SSA the contribution of the informal sector to non-agriculture GDP is about 41 per cent. Hence, the efficiency of microenterprises matters in determining overall economic performance and poverty reduction.

Despite their potential to improve economic growth, micro and small enterprises (MSEs) in developing countries fall short of expectations. They produce largely for the low-income group and employ lower levels of techniques. Many microenterprises are of the self-employed type with low rates of graduation into higher size categories and their innovative activities are limited (Kiggundu 2002). This is largely due to the harsh environment they operate in. Unreliable enforcement of contracts, excessive regulatory and administrative requirements, limited access to finance, and inadequate infrastructure services all impose disproportionately high transaction costs on MSEs for doing business generally, and for innovative activity in particular (Ernst 2004).

The promotion of MSEs is becoming a popular development tool. Accordingly, governments and donors in developing countries have shown increasing interest in promoting innovation and entrepreneurship. They have initiated various support programmes with the aim to improve MSEs' competitiveness through enhancing technology and innovation capabilities, such as upgrading product quality, improving design and packaging, and training to improve competitiveness (Pyke 1994). The notion is that innovation is essential for MSEs to become and remain competitive, move to higher return activities, and to grow and graduate to small and medium-sized enterprise status, thus creating new employment opportunities (Ernst 2004). Improving competitiveness is even more crucial in the context of liberalization and increasing integration into the world market. Lack of adaptation and upgrading spells defeat, while firms that keep up or even initiate their own original improvements can be expected to perform well (Romijn 2002).

The efficacy of such interventions, however, depends on identifying key factors that foster or inhibit innovation by MSEs and target the potentially successful entrepreneurs. Small business entrepreneurs are hardly homogeneous in objectives and capabilities. Many are self-employed while others have high vigour to innovate and grow. They also differ in terms of socioeconomic background and access to resources such as financial capital. The types of activities they are in are also diverse. What types of entrepreneurs/firms are more likely to engage in innovative activity? Do innovators grow faster and create more jobs than non-innovators as it is claimed? Understanding the attributes of innovators and their impact on employment is crucial in order to formulate effective policies.

Despite the high profile of the issue in current policy formulations in Africa, there is little empirical evidence on innovativeness and its impact on firm performance in MSEs. The existing few studies in Africa mainly examine the determinants of innovative activity and attributes of innovativeness (for example, van Dijk 2002; Oyelaran-Oyeyinka 2006; Robson, Haugh, and Obeng 2008). Van Dijk (2002) examines the importance of enterprise clusters and cooperation on innovation in the informal sector in Ghana, Burkina Faso, and Zimbabwe. Oyelaran-Oyeyinka (2006) analyses the impact of interfirm collaboration on innovation in Kenya, Nigeria, and Zimbabwe using 200 manufacturing firms. Robson, Haugh, and Obeng (2008) investigate the determinants of innovation in Ghanaian small enterprises that employ between four and 50 workers. The lack of empirical evidence is even more apparent when it comes to the effect of innovation activity on firm growth. Mahemba and de Bruijn (2003) report only weak association between innovativeness in small firms and growth in the Tanzanian manufacturing sector. Thus, the relationship between innovativeness and small firms' growth has not yet been empirically confirmed in Africa.

In this chapter we seek to address two interrelated issues with regard to African MSEs: the determinants of innovative activity and whether innovative enterprises grow faster than non-innovating enterprises. We use a large set of microenterprise survey data from Ethiopia that comprise 1,000 observations from six selected major towns including the capital city Addis Ababa. As in other developing countries, in Ethiopia, the informal sector plays a significant role in the economy. According to the 1999 survey by the Central Statistical Agency (CSA), the urban informal sector comprises about 50.6 per cent of total urban employment of the 2.88 million persons.[1] Female employment accounts for about 58 per cent of the employment in the informal sector.

Recognizing the significance of this sector, the Ethiopian government issued a National Micro and Small Enterprises Strategy in 1997 and established the Federal Micro and Small Enterprises Development Agency in 1998. The country's industrial policy in 2003 and the poverty reduction strategy in 2006 have singled out MSEs as major instruments to create a productive and vibrant private sector and reduce poverty among urban dwellers. These documents reiterate the importance of MSEs' promotion through the provision of finance, training, and infrastructure services among others. However, in our data only a few enterprises (no more than 8 per cent) reported that they have received some support from government or NGOs. This implies that the innovation activity of the microenterprises is expected to be a result of the decision of the owner. Our analysis will, therefore, emphasize the entrepreneurs' behaviour and the resources available to the enterprises as major determinants of innovativeness and firm growth.

This chapter contributes to the thin literature on innovations in African MSEs in the following ways: first, it analyses not only the determinants of innovation but also the impact of innovation on firm employment growth. Second, it exclusively relies on the lower bottom of size category, firms with ten and fewer workers usually termed as microenterprises. By doing so, this study tries to address the bias that might arise from pooling a heterogeneous group in the previous studies as a result of broader definition of small enterprises, i.e. up to 100 or so workers. Third, unlike most previous studies it covers not only manufacturing but also other major sectors such as services and trade.

6.2 Data

The data source of this study is a survey of a 1000 microenterprises with ten or less workers conducted in 2003 by the Ethiopian Development Research Institute. The survey was done in six selected major towns: Addis Ababa, Awassa, Bahir Dar, Jimma, Mekelle, and Nazreth. A total sample of 974 enterprises was interviewed; 25 per cent of the enterprises were from Addis Ababa and about 15 per cent from each of the other cities.[2] Table 6.1 gives

Table 6.1 Characteristics of microenterprises and owners

		N	%
Sector	Trade	439	45
	Service	349	36
	Manufacturing	186	19
Firm size category	1 worker	172	18
	2–4 workers	493	51
	5–10 workers	302	31
	>10 workers	6	0.6
Firm age category	<= 5 years	439	45
	6–12 years	347	36
	13–29 years	141	14
	above 29 years	47	5
Owner gender	Female	226	22.3
	Male	722	74
Owner age group	<=25	185	19
	26–35	392	40
	36–50	247	25
	50 and above	150	15
Owner education	Illiterate	113	12
	Elementary	287	29
	Some high-school	119	12
	High-school complete	310	32
	Some college years	145	15
	Has vocational training	182	19

the distribution of the enterprises and characteristics of the owners in our sample. The enterprises cover a wide variety of non-agricultural activities like trade, services, and manufacturing. The majority of them are engaged in trade and services, constituting 45 per cent and 36 per cent, respectively. Manufacturing is also an important component of the microenterprises (19 per cent) mainly covering production activities such as wood and metal work, bakeries, and tailors.

Measuring the number of workers as the sum of working owners, paid and unpaid workers in 2002 (one year before the survey), 69 per cent of the businesses have less than five workers of which one-worker establishments constitute about 18 per cent.[3] Firms that have five to ten workers account for 30 per cent. Most of the enterprises are young, whereby 45 per cent of them are 5 years old or younger, and 36 per cent are 6 to 12 years old. Male-headed businesses account for 74 per cent, while only 22 per cent are female-headed. The female-headed businesses tend to concentrate on activities like retail trading, beauty salon, bars and restaurants, and local drink brewing. The majority of the owners are young, 59 per cent are under 35 years old. The survey instrument also includes questions about the owners' educational attainments: 32 per cent of the owners have completed high-school and

Table 6.2 Type of innovative activities by sector

If yes, what improvements/ changes?	Broad category of innovation	Frequency by sector			
		Trade	Service	Manufacturing	All
Improve quality product/ service	Product/ service	20	20	6	46
Provide new products/service	Product/ service	9	13	3	25
Better design	Product/ service	9	12	18	39
Increase variety of products/ services	Product/ service	12	5		17
Install additional machinery	Process	1	4	9	14
Introduce modern machinery	Process	6	7	15	28
Additional business premises/ house	Process	5	4	2	11
Additional utensils/furniture/ equipment	Process	15	16	3	34
Renovation	Process	13	14	3	30
Improved production capacity	Process	12	3	5	20
More advertisement	Marketing	11	3	4	18
Shorten delivery time	Marketing	1	3	1	5
Discount	Marketing		2		2
Accounting system	Organization & skill			1	1
Managerial skill	Organization & skill	3	2		5
Hired skilled worker	Organization & skill		1		1
Skill improvement	Organization & skill	2	7	9	18
Additional business partner	Organization & skill		1		1
Expanded the business	Organization & skill	12	1	5	18
Total number of firms that reported recent important improvements/changes		131	118	84	333
Total number of firms in the survey		439	349	186	974
Percentage of firms that reported recent important improvements/changes		0.30	0.34	0.45	0.34

15 per cent have some years of college, while 12 per cent are illiterate; about 15 per cent of the owners have also reported that they had vocational training.

Our innovation indicator is a dichotomous variable that takes value one if the respondent replied 'yes' to the question 'Did you make an important improvement/change to your product/service recently?'. As shown in the

Table 6.2, about 34 per cent of the enterprises said 'yes'. Those responding 'yes' were then asked to disclose what type of improvement was mainly involved. The lists of activities show about 20 types. We categorized them into main type of innovative activities, such as product/service innovation (providing new/quality/better design or an increasing variety of products), process innovation (machinery investment, improving or increasing business premises, furniture, and equipment), organizational and skill improvement (improving the skill of workers and managers), marketing (more advertisement, shorter delivery time). These activities are more or less incremental and consistent with the observation made by van Dijk and Sandee (2002) on innovation in African small firms.[4]

The magnitude of innovativeness differs slightly by sector. The manufacturing sector has a higher propensity to innovate than the service and trade sectors as shown by the percentage of firms that reported they had improved their products/services in each sector. But more importantly the sectors differ in the type of innovative activity. The manufacturing sector is distinctive from trade and service in this context. The innovation activities in the manufacturing sector are machinery improvement (investment), better design, or skill improvements. Service and trade sectors, on the other hand, tend to concentrate on the improvement of business premises, provision of quality and variety of products/service, and marketing.

Looking at the association between the innovative indicator and other variables defining the characteristics of the owners and their perceptions might give some guidance to the empirical analysis of the determinants of innovation. The survey instrument includes questions about owner perceptions about innovation, particularly with regard to the relative status of his/her business in terms of innovation. For example, how do you compare your main product/services with that of your competitors in terms of quality material and model/design? How do you characterize the enterprise's machinery/equipment? The survey instruments also include business environment perception and a variety of owner/firm attributes.

Table 6.3 presents the correlation coefficient between the innovation indicator and other variables. Stars represent significance at 5 per cent or better. The innovative indicator is positively associated with the owner perceptions such that his/her business has better quality material and design than the competitors and uses advanced machinery. The innovators subgroup perception on business environment is also more optimistic than the non-innovators group. The innovation's indicator is positively associated (and significantly) with current size, employment growth, investment, revenue increase, having no market problem and plans to expand the business in the future. Among the demographic characteristics (owner age, gender, and marital status), only gender is found to be significantly associated with

Table 6.3 Correlation coefficients between innovativeness and owner attributes

Correlation with improved/changed products/services recently			
Variable	Correlation coefficient	Variable	Correlation coefficient
1) Perception of innovativeness		3) Demography	
Higher quality than competitors	0.1795*	Owner female	−0.0643*
Higher design than competitors	0.0744*	Owner age	−0.0174
Advanced machinery	0.0992*	Married	−0.0025
2) Perception of business environment		4) Education/ experience	
Revenue increased	0.0930*	Illiterate	−0.0750*
Have no market problem	0.1500*	Elementary	−0.0249
Have plan to expand	0.1610*	High-school	0.0127
Need credit but did not get any from formal source	−0.0029	High-school complete	0.0087
Employment change 2001–3	0.1035*	Some college	0.0770*
Current size category	0.1383*	Received vocational training	0.1465*
Cumulative investment 2001–3	0.0866*	Experience in business	0.0211

Note: *denote significance at 5% or better.

innovation activity. This indicates that female owners are less likely to engage in innovative activity than male owners. We have also tested for association between innovation activity and owner education and experience. Owners with vocational training and some college years are more likely to innovate, while illiterate owners are less likely do so.

In the remaining part of this section we examine the growth pattern of firms and particularly show how many of the businesses are able to transit from one to another size class in either direction. Based on the initial size category when

Table 6.4 Transition across different size categories

		Size category 2003				
		One worker	2–4 workers	5–10 workers	>10 workers	total
Size category at start	One worker	150 (0.56)	101 (0.38)	18 (0.07)		269
	2–4 workers	21 (0.04)	375 (0.71)	133 (0.25)		529
	5–10 workers	2 (0.01)	16 (0.10)	146 (0.87)	4 (0.02)	168
	>10 workers		1 (0.13)	5 (0.63)	2 (0.25)	8
	total	173	493	302	6	974

Note: numbers in parentheses represent the ratio of number of firms in the given cell in 2003 to the number of firms in that size group at initial year.

starting the business we calculated the percentage of firms that was able to transit to another size category at the end of the period. Table 6.4 gives the transition of the establishments across size classes from the start to the current period. The four categories are one-worker, two to four workers, five to ten and above ten workers. In general most of the firms, i.e. about 69 per cent of the establishments did not change their size category in the given period.

However, a large number of one-worker businesses have increased their size, i.e. about 38 per cent of one-worker establishments at start grew to the next size class (two to four) and 7 per cent of them to the higher size class (five to ten) at the end of the period. This contrasts with the prevailing notion that for a one-person firm self-employment may be an alternative to working for another firm, as a consequence of which the person may have no desire to grow or may lack the know-how to do so (Biggs and Srivastava 1996; Mead and Liedholm 1998).

About 25 per cent of the microfirms with two to four workers at the beginning of the period have also evolved to the small firm category with five to ten workers at the end of the period. A downsizing of about 11 per cent has also been observed among the small firms.

6.3 Determinants of innovation activities

6.3.1 *The framework*

Innovation was seen as a breakthrough or *radical change* characterized by marketing and technological discontinuity and primarily produced by large firms and concentrated markets as argued in the early works of Schumpeter. However, innovation is rarely a dramatic breakthrough, rather a small improvement in a new process or product—'incremental innovation' (Blaug 1999). The realization of economic benefits from 'radical' innovations in most cases requires a series of incremental improvements. Thus, the bulk of economic benefits come from incremental innovations and improvements (Fagerberg 2005). Broadly defined innovative activities include introducing new products/services, new design, and improved quality of products/services, installed new equipment, changed sales methods, and improved working conditions. An innovation in a small enterprise in the developing countries context is largely an adoption of a product, process, or method that has already been adopted elsewhere but is new to the firm and not necessarily new to the world, region, country, or industry (van Dijk 2002).

What factors determine innovativeness and innovation activities in small firms? Hyvärinen (1990) identifies three broad (sometimes overlapping) determinants: entrepreneurial attributes, firm-level resources, and the environment in which the firm operates.

ENTREPRENEURIAL CHARACTER

In small businesses decision-making is concentrated in the hands of the owner/manager (Dyer and Handler 1994). Innovation activities of individuals (owner/manager) form an important part of the innovation activities in small enterprises. Thus, the smaller the enterprise, the nearer its innovative behaviour is to that of an individual's (the owner/manager) behaviour (Hyvärinen 1990). Any attempt to investigate the innovation needs to consider analysis of the characteristics of the entrepreneur (Donckels and Fröhlich 1991; Hausman 2005). Rogers (1995) summarizes characteristics of innovative entrepreneurs into three headings: socioeconomic status, personality, and communication behaviour of which education, social status, age, attitude toward risk and science, and density of social network in which the individual participates are among the long list of variables.

Various empirical studies have tested the effect of human capital and demographic factors on innovation. Khan and Manopichetwattana (1989) and Hausman (2005) showed that in the USA innovative firms are led by more educated executives or owners. In Ghana, Robson, Haugh, and Obeng (2008) found that educated owners are more likely to innovate. The experience of owners (level of skill and knowledge) is also an important factor and has been found to affect innovation activities (Hausman 2005). Mahemba and de Bruijn (2003) and Robson, Haugh, and Obeng (2008) have also shown that training of workers is associated with higher innovation. Khan and Manopichetwattana (1989) found that firms led by on average younger owners, are pro-active, risk-taking, and more innovative. So far, the relation between the owner's gender and innovation activities has not been empirically established. In the entrepreneur literature, however, there is some evidence showing that female-headed firms grow slower than male-headed ones (Liedholm and Mead 1993; McPherson 1996).

FIRM-LEVEL FACTORS

Innovation activity occurs at firm-level and the firm is a central actor in processes of technological change (Romijn 2002). In the empirical literature these resources are captured by firm size, age, access to finance, and network. The relation between firm size and innovative activity is a longstanding debate since the work of Schumpeter (1939).[5] However, the empirical results so far are not conclusive (Nooteboom 1994; Ernst 2004). We are not pursuing this debate as our data covers only the lower segment of firm size with ten and fewer employees. But the size of a firm could still impact on innovative activities even within the microenterprises, capturing differences in access to resources. Rogers (1995) indicated that early adopters are wealthier and have large-sized units (farms, schools, companies, and so on).

Innovative spirit could be associated with the age of a firm in the sense that small firms have higher innovative capacity in the first stage of a life cycle. In contrast, firm age could also represent accumulated resources, market knowledge, and developed networks so that older firms would be more likely to be involved in innovation activities. The empirical evidence in Africa so far is mixed. Deraniyagala and Semboja (1999) and Wignaraja (2002) found supporting evidence of a positive relation between firm age and innovation and technological capability. Robson, Haugh, and Obeng (2008), however, found no significant relation between firm age and innovation.

Innovation activities take place more easily in clusters and networks (van Dijk and Sandee 2002). Effective networks that comprise lateral and vertical linkages raise capacity for each node in the network by increasing exposure to ideas and opportunities. They also reduce the transaction of developing and adopting innovations (Ernst 2004). A voluminous empirical literature supports the role of clusters and networks on innovation in Africa (Sverrisson 1997; Chipika and Wilson 2006; Oyelaran-Oyeyinka 2006). Unfortunately, we do not have a good approximation of network in our data and thus have not included a network variable in the empirical analysis. We believe that if such effects exist, then firm age might partly capture the impact of the network.

Based on this brief survey, the descriptive analysis in the previous section, and availability of data we propose to test the following hypotheses:

Hypothesis 1: Entrepreneurs with more formal education, technical/vocational training, and/or longer previous experience are more likely to undertake innovative activity.

Hypothesis 2: Younger entrepreneurs are more likely to undertake innovative activity.

Hypothesis 3: Male entrepreneurs are more likely to undertake innovative activity than female entrepreneurs.

Hypothesis 4: Larger firms are more likely to innovate than smaller firms.

Hypothesis 5: Older firms are more likely to innovate than younger firms.

6.3.2 *Estimation and results*

Discriminant analysis and logistic regression are widely applied for identifying the attributes of innovative and non-innovative entrepreneurs/firms (for example, Ostlund 1974; Kim and Kim 1985; Oyelaran-Oyeyinka 2006; Moreno and Casillas 2007; Koellinger 2008; Robson, Haugh, and Obeng 2008). Discriminant analysis, however, is based on a number of assumptions that sometimes are difficult to justify. It requires assumptions such as normal distribution, linear and homoscedastic relationships, untrancated interval or near interval data, proper model specification, and whether the dependent variable is a true dichotomy, among others. In contrast, logistic regression requires no assumptions regarding the distribution of the explanatory variables. It is relatively robust, flexible, and easily used, and lends itself to a meaningful interpretation. Logistic regression is preferred when data are not normal in distribution or group sizes are very unequal (Garson 1998/2008; Pohar, Blas, and Turk 2004). Thus, in this chapter we applied logistic regression to test the above hypotheses about the determinants of innovative activity.

The dependent variable 'innovativeness' is defined here as a categorical variable indicating that the firm made important improvements/changes to its product/service recently. The independent variables include firm size, firm age, and owner attributes such as age, gender, previous business experience, general education, and vocational training. The size of the firm is measured by the current number of employees. Firm age is the number of years since start-up in logarithmic form. Business experience represents the experience of the owner in business before starting the current business measured in number of months and in logarithmic form. The owner's age is also in logarithmic form. We made a distinction between general education and vocational technical training. General education is represented by a high-school certificate and some years of college education. This means that the owner who did not complete high-school and the illiterate serve as a control category. Vocational training is a dummy capturing if the owner had access to vocational training before or after starting the business. Two sectoral dummies representing manufacturing and service are included, with trade as control category. Regional difference is also controlled for with Addis Ababa as a reference city.

Table 6.5 reports the logit estimation results. The size of the firm is positive and highly significant. This means larger firms are more likely to participate in innovation activity. The positive effect of size indicates the resource advantage of larger firms over smaller ones and it is consistent with the theory of a resource-based view and previous studies (for example, Robson, Haugh, and Obeng 2008). Firm age is also positive and significant. But when we include age square into the model we find a non-linear relationship (i.e. positive at the first level and negative squared term, both significant) between innovation and firm age (see Table 6.5, column (2)). The concave relationship between firm age and innovativeness suggests that innovative activity increases at an early age but tends to decline beyond a certain age. The positive relation between age and innovation activity at an early age might be due to accumulated business experience and market knowledge. However, this advantage might not last long. The manufacturing dummy is positive and significant

Table 6.5 Determinants of innovation activities: logit estimation

	Dependent variable dummy for innovation activity		
	(1)	(2)	(3)
Firm attributes Δ employment (2001–3)			0.012
			(0.057)
Current employment	0.153***	0.155***	0.153***
	(0.0340)	(0.034)	(0.036)
log(firm age)	0.194**	0.682***	0.687***
	(0.0886)	(0.198)	(0.199)
log(firm age)^2		−0.100***	−0.100***
		(0.037)	(0.037)
Owner attributes			
log(owner age)	−0.706***	−0.851***	−0.851***
	(0.268)	(0.274)	(0.274)
Owner female	−0.388**	−0.383**	−0.381**
	(0.184)	(0.185)	(0.185)
log(previous business experience)	−0.00943	−0.002	−0.002
	(0.0430)	(0.043)	(0.043)
Owner high-school complete and above	−0.0505	−0.032	−0.032
	(0.168)	(0.169)	(0.169)
Owner has vocational training	0.871***	0.872***	0.870***
	(0.204)	(0.205)	(0.205)
Sector (control category trade)			
Manufacturing	0.678***	0.687***	0.682***
	(0.202)	(0.203)	(0.204)
Service	0.223	0.265	0.265
	(0.172)	(0.174)	(0.174)
Constant	0.262	0.266	0.264
	(0.931)	(0.933)	(0.938)
Region controlled	Yes	Yes	Yes
Observations	966	966	966
Pseudo R2	0.13	0.135	0.136

Note: figures in parentheses are standard errors, and ***p<0.01, **p<0.05, *p<0.

suggesting that manufacturing firms are more likely to engage in innovation activity compared.

The coefficient of female owners is negative and significant suggesting female-owned enterprises are less likely to innovate in contrast to those owned by male entrepreneurs. This is usually explained by the fact that women owners are more family-oriented and interested in the long-term stability of the business, thus tend to take less risk (Brush 1992), and as a result are less likely to engage in innovation activity. Women entrepreneurs also face more operational and strategic impediments compared to male entrepreneurs in their entrepreneurial pursuit (Rutashobya 2001). Owner age is also negative and significant: the older the age of the entrepreneur the less likely they are to innovate. This is consistent with the Khan and

Manopichetwattana (1989) finding and might be explained by the fact that older entrepreneurs are also more risk-averse than younger entrepreneurs. Among the human capital variables only vocational training is found to affect innovation activity positively and significantly. Neither general education nor previous experience has significant effects. This lends support to the belief that vocational (technical) training is more important than general education in promoting entrepreneurship.

In the last column of Table 6.5, we introduce employment growth measured by cumulative change of employment in the firm between 2001 and 2003. This was to test if there is any causation from growth to innovation. However, firm growth is not significant while the effect of other variables remains the same. Robson, Haugh, and Obeng (2008) have also found no clear relation from firm growth to innovation in their estimation of innovation equation that includes firm growth. The effect might be the other way round, i.e. from innovation to firm growth.

6.4 Innovativeness and firm growth

6.4.1 *The framework*

The second objective of this study is to investigate if innovative firms grow faster than non-innovative firms, in other words, if innovators create more jobs. Theoretically, new technologies and processes are associated with a better utilization of resources, higher quality of routine tasks and higher productivity. Companies that use innovative technologies and processes can often offer qualitatively superior and/or cheaper products, thereby enjoying higher growth potential (Minitti, Bygrave, and Autio 2006). McDaniel (2000) also argues that those firms successfully mastering the timing and placement of innovation development and innovation implementation in their respective industries will be set to lead in profit markets' share and industry dominance. In the context of new firms Geroski (1995) argues that the growth and survival prospects of firms will depend on their ability to learn about their environment, and to link changes in their strategy choices to the changing configuration of that environment.

Although there are also counter-arguments that innovation might replace employment most empirical studies in the developed world show that innovative firms are more likely to grow in terms of market shares, profits, or employment, regardless of industry, size, or other characteristics. Mansfield (1962) compared average annual growth rates of innovators and non-innovators with comparable initial size in US manufacturing and found that the growth rate of innovators is about 4–13 percentage points higher than that of the control group. Jones-Evans et al. (1996) found that technologically innovative small and medium-sized enterprises (SMEs) in the UK have growth rates above the average regarding assets and expropriations. Moreover, such companies tend to have low bankruptcy rates. Koellinger (2008) examines the relationship between the usage of internet-based technologies, different types of innovation, and performance at the firm-level based on a sample of 7302 European firms. He found that all considered types of innovation, including internet enabled and non-internet enabled product or process innovations, are positively associated with turnover and employment growth but innovative activity is not necessarily associated with higher profitability. Cho and Pucik (2005) conducted research using data from the Fortune Reputation Survey and the Research Insight Global Vantage. They found a significant relationship between innovativeness and firm growth and profitability. There is not much empirical work in Africa on the impact of innovation on firm growth. Mahemba and de Bruijn (2003) found no clear relation between innovativeness and firm growth in small manufacturing firms in Tanzania.

Besides innovative activity a number of other factors could also influence firm growth. The stochastic theory of Gibrat's law relates size distribution and firm growth and argues growth is independent on size. However, the growing empirical literature shows the contrary, i.e. a negative relationship between growth and firm size (Evans 1987; Dunne and Hughes 1994; Bigsten and Gebreeyesus 2007). The inverse relation between firm growth and size can be explained through the availability of slack resources suggested by Penrose (1959). Such idle resources arise as a consequence of their indivisibility. The extent to which a firm can employ the most advantageous division of labour depends on the scale of its operations; the smaller its output the less resources can be used in a specialized manner. The smaller the firm, the greater the indivisibility of resources and availability of slack resources, thus the higher the incentive to expand.

In a life cycle theory of the firm, Jovanovic's (1982) passive learning model predicts a negative relation between firm age and growth. This is supported by several empirical studies (for example, Evans 1987; Dunne and Hughes 1994; McPherson 1996; Bigsten and Gebreeyesus 2007). The active learning model by Ericson and Pakes (1995), on the other hand, argues that firms which invest in R&D and human and physical capital will be more efficient and grow faster.

Lack of access to financial resources hinders firms from growing to their optimal size (Holtz-Eakin, Joulfaian, and Rosen 1994; Elston 2002; Cabral and Mata 2003). Micro and small enterprises are more likely to face liquidity problems as they are considered expensive to be served thus less attractive to the formal banks. Lack of finance is the most often mentioned complaint among entrepreneurs in Africa (for example, Biggs and Srivastava 1996; Bigsten et al. 2003). Other studies have also related firm growth to entrepreneurial attributes such as owner age, and gender (for example, Davidson 1991; Liedholm and Mead 1993; McPherson 1996; Davidson and Honig 2003).

In what follows we will analyse the impact of innovation on firm growth controlling for size, age, credit constraints, and owner characteristics. In modelling the relation between innovation and firm growth, we start with the Evans (1987) firm growth equation that relates growth with initial size and age, but augmenting it with innovation indicators and other variables.

$$(\ln S_{t+n} - \ln S_t)/n = \beta_0 + \beta_1 \ln (S_t) + \beta_2 \ln A_t + \beta_3 INN + \sum \gamma_i X_i + u_t \qquad (1)$$

where S_{t+n} and S_t represent employment size at year $t+n$ and year t respectively (in our case the logarithm of size in 2003 and logarithm of size pre-2001), n denotes the number of years the firm operated between year t and $t+n$, A denotes firm age, INN innovation indicator (a dichotomous variable), and X indicates other control variables (financial constraint and owner education, experience, and age), and u is the log-normally distributed errors term with mean zero. Credit constraint is defined as =1 if the firm reported that it needs credit but is unable to borrow due to different reasons, and zero otherwise. The other explanatory variables were defined in the previous section.

Size and age often have a non-linear relationship with firm growth. A non-linear relationship between size and growth might not make sense in a small range of size such as the data we have of firms with ten and less workers. Nonetheless, the range of age of the firms is very wide from 1 to over 60 years old (see Table 6.1). Thus, we estimated a non-linear relationship between age and growth by introducing the square of log age of the firm into the equation.

In the literature there are different measures of firm performance, such as growth in sales, profits, market share, assets, and employment. In this analysis we confined ourselves to employment growth basically due to the absence of sufficient sales and assets variables in our data. The dependent variable is average annual employment calculated as the logarithmic difference between the 2003 and pre-2001 employment and divided by the number of years the firm operated in the sample period.

6.4.2 *Estimation and results*

A common statistical problem in such a model is sample censoring. One source of the sample censoring is firm exit. A failure is common among small firms. Small firms that have slow or negative growth are more likely to exit than larger firms. Thus the proportional rate of growth conditional on survival will be small for larger firms. Ignoring this problem might result in a downward bias estimate in the relationship between growth and size of firms, although, this bias turns out to be insignificant in many previous studies (Evans 1987; Hall 1987; McPherson 1996). Moreover, our sample is also censored from above, i.e. the observations included are only those businesses with ten and less workers. This means firms that increase their workers' number above ten are also excluded from the data and this might reinforce the likely spurious negative relation between growth and initial size. Avoiding all these possible biases might be difficult in a cross-section data such as the one we have. However, if such bias exists and is important then we should find different results when estimating the growth equation for the sub-sample of smaller size, for example firms with one to four workers at the initial period (pre-2001). Hence, as a robustness check we will take a sub-sample of micro-firms with one to four employees and estimate the same model.

Estimating the above equation with ordinary least squares (OLS) might lead to inconsistent results if one or more of the right hand side variables are not exogenous to the model. As we have shown in the previous section innovation is determined by many of the variables in the model, such as entrepreneurial characteristics. We performed a test for endogeneity of the innovation indicator using an augmented regression test and found that the innovation indicator is correlated with the error term which makes OLS results inconsistent. There are different methods in controlling the endogeneity problem: simultaneous equation, fixed effect, instrumental variable (IV) method and so forth. Given that our data is cross-section we use the IV method, specifically the two-stage least square method (2SLS). The disadvantage of the IV method is that identifying a proper instrument is not easy. The requirement for proper instrument is that it should be correlated with the instrumented (i.e. included endogenous regressors) and at the same time uncorrelated with the error term.

Based on some experimentation of the data vocational training and owner gender are found to satisfy the requirement for proper instrument. The first stage regression and test results are reported in Appendix table 6.1. First, the instruments are highly correlated with the innovation variable. F-tests that the correlations with gender and vocational training equal zero are also rejected. However, simply having an F-statistic that is significant at the typical 5 per cent or 10 per cent level is not sufficient. Stock, Wright, and Yogo (2002) suggest that the F-statistic should exceed ten for inference based on the 2SLS estimator to be reliable when there is one endogenous regressor.[6] The F-test statistic from the first regression is 12.14 and exceeds ten—the rule of thumb (see Appendix table 6.1). Hence, the instruments are strongly correlated with the endogenous regressor. We have also conducted a test overidentifying restriction (see Table 6.5).[7] The Sargan-statistic of overidentification restriction cannot be rejected implying the instruments are valid. Confirming that our instruments are valid we now proceed to discuss the results.

Table 6.6 Firm growth and innovation activities

Variables	Average annual employment growth (2001–3)			Net change of employment (2001–3)	
	(1)	(2)	(3)	(4)	(5)
Innovation dummy	0.166**	0.168**	0.193**	1.223**	1.237**
	(0.0803)	(0.0809)	(0.0909)	(0.626)	(0.637)
ln(employment) $_{pre-2001}$	−0.0995***		−0.112***	−0.224*** [a]	
	(0.0110)		(0.0161)	(0.0244)	
Size category (2–4 workers)		−0.115***			−0.519***
		(0.0152)			(0.116)
Size category (>=5 workers)		−0.183***			−1.184***
		(0.0212)			(0.165)
ln(age)	−0.0861***	−0.0856***	−0.104***	−0.301**	−0.298**
	(0.0169)	(0.0170)	(0.0208)	(0.129)	(0.130)
ln(age)^2	0.0128***	0.0122***	0.0146***	0.0494**	0.0427*
	(0.00304)	(0.00303)	(0.00369)	(0.0232)	(0.0234)
ln(previous experience years)	0.00998***	0.00944***	0.0136***	0.0575**	0.0490*
	(0.00332)	(0.00334)	(0.00420)	(0.0251)	(0.0257)
ln(owner age)	0.0814***	0.0740***	0.0850***	0.742***	0.678***
	(0.0236)	(0.0235)	(0.0282)	(0.178)	(0.182)
Credit constrained (dummy)	−0.0561***	−0.0544***	−0.0663***	−0.499***	−0.482***
	(0.0131)	(0.0132)	(0.0171)	(0.0991)	(0.101)
Owner high-school cert. and above	0.0285**	0.0266**	0.0312*	0.180*	0.173*
	(0.0129)	(0.0129)	(0.0165)	(0.0982)	(0.100)
Manufacturing	0.0361*	0.0312	0.0334	0.550***	0.530***
	(0.0206)	(0.0207)	(0.0273)	(0.156)	(0.160)
Service	0.0268*	0.0206	0.0231	0.291**	0.235***
	(0.0138)	(0.0140)	(0.0188)	(0.106)	(0.108)
Constant	−0.0847	−0.0405	−0.0826	−1.502**	−1.365**
	(0.0827)	(0.0828)	(0.0988)	(0.620)	(0.637)
Regions controlled	Yes	Yes	Yes	Yes	Yes
Observations	941	941	715	965	965
Overidentification test					
Sargan statistics (Chi-square)	0.0687	0.06679	0.0833	0.841	0.7817
p-value	0.793	0.796	0.772	0.358	0.376

Note: Standard errors in parentheses, ***p<0.01, **p<0.05, *p<0.1
[a] the size variable is not in log format.

Table 6.6 reports the 2SLS estimation results.[8] The first three columns use annual average growth of employment as a dependent variable. Column (1) is the basic model and makes use of the full sample. In column (2) we introduce dummies for size categories (one-worker firm as reference category) instead of absolute size. Column (3) reports the estimation results of a sub-sample of smaller firms (i.e. firms with only one to four employees at the start of the sample period) as a robustness check for a possible bias from sample censoring. The last two columns (4) and (5) report estimation results for which the dependent variable is net change of employment between 2001 and 2003 instead of growth. This is another check for the robustness of our results from a bias that might arise from growth calculation. The common growth calculations tend to give overstated growth for firms with very small initial size, due to a smaller denominator.[9]

Our main interest in this sub-section is to test if innovation impacts firm growth. In all the estimations, the innovativeness indicator is positive and significant at the 5 per cent level. This means the growth and innovativeness relation is robust to the different specifications of the dependent variable, when taking sub-samples of smaller firms, and control of various variables that impact firm growth. The results confirm our hypothesis that innovative firms grow faster than non-innovators. Thus, innovative activity predicts higher job creation. Based on the growth estimation results, *ceteris paribus* innovators on average grow 16.6 to 19.3 per cent more than non-innovators (depending on the specification).

Other variables are also found to affect growth of employment in MSEs. All results including the sub-sample with one to four workers give a negative and highly significant coefficient of initial size (i.e. pre-2001). The estimation with size category also gives a negative coefficient of higher size groups in contrast to the control group—a one-worker size group. These results suggest that smaller firms grow faster than larger firms and this is robust to sample censoring and other specifications. However, it is difficult to control all sources of bias with the type of the data we have and these interpretations should be taken with caution.

In all the estimations we found a convex relationship between age and growth of employment with the first level taking a negative sign and the squared term positive, both significant. This means firm age is related with growth negatively, but the negative relation diminishes with age. The negative segment captures evidence of a learning process that was proposed by Jovanovic (1982) whereby as a firm ages and grows more confident about its costs, the mean and variance of its growth rate should decrease. This is also consistent with the previous findings. Bigsten and Gebreeyesus (2007) found a convex relation in the Ethiopian manufacturing sector. Evans (1987) reported that firm growth decreases with age for younger firms but is roughly independent of age for older firms in US manufacturing.

Credit constraint is highly significant and negatively related with firm growth. It is obviously an important factor hindering firm growth given that about 85 per cent of the firms in our sample have never received credit from the formal market, like banks and microfinance institutions. They largely depend on the informal network of relatives and friends, and trade credit. We have also found a positive and significant impact from owner age, previous experience, and education on firm growth. These suggest that entrepreneurs with more education and previous experience and those who are older are more likely to grow. In all estimations, we have included sector dummies, i.e. manufacturing and service but found less conclusive results. Both the manufacturing and service have positive coefficients but not always significant at standard levels.

6.5 Summary and conclusions

The aim of this chapter was to address two prominent issues in the MSEs' development in Africa. The first is to show the factors that foster or constrain innovation and the second to examine if innovative enterprises create more jobs than non-innovators. We estimated separate models of innovation and growth determinants. We used a logit estimation method for the innovation model. In the growth equation we applied the IV method to address the endogeneity of innovation in the model. Appendix table 6.2 summarizes the signs and significance levels of the variables in both models.

Innovation activity is related with a number of firm and entrepreneur attributes. Current size is related positively with innovation activity. This means the larger the size the more likely the firm is involved in innovative activity. Resource advantage could explain why larger firms are more innovative than smaller firms. This is consistent with most previous studies. We found a non-linear (concave) relationship between firm age and innovation activities. Innovative activity increases at early age, but tends to decline beyond a certain age. Our interpretation of the results is that the positive relation between firm age and innovation activity at early age might be due to accumulated business experience and market knowledge. However, this advantage might not last long. For example the innovative spirit of firms might decline with age.

Among the human capital variables vocational training is found to have a strong effect on the innovation activity. Unlike other studies (for example, Hausman 2005), neither general education (measured by high-school certificate and above), nor previous business experience are affecting innovation in our data. This gives support to the notion that technical skill is more important than general education in promoting entrepreneurship and innovation.

Female-owned firms are less likely to be involved in innovation activity. This might be explained by the fact that female entrepreneurs are risk-averse. Previous studies indicated that female entrepreneurs are more family-oriented and thus less interested in the expansion of their businesses (Brush 1992). In developing countries female entrepreneurs also face more constraints compared to male entrepreneurs in their entrepreneurial pursuits. We have also found a negative relation between innovativeness and owner age suggesting that younger owners are more likely to innovate than older ones. This is usually an indication of the owners' risk attitude.

The main contribution of this study to the MSEs literature in Africa is its extension in examining the effect of innovativeness on firm growth while controlling a range of other potential variables that could have an effect on firm growth (e.g. size, age, financial constraint). We found strong evidence that innovators are more likely to grow than non-innovators. On the other hand, we found no evidence of the reverse causation (i.e. from growth to innovativeness). This supports the claim that innovations lead to the expansion of businesses and creation of more jobs. A focus on promoting innovation and technological capability will, therefore, pay off not only through increasing MSEs' competitiveness but also through their ability to create more jobs.

APPENDIX

Appendix table 6.1 Results of the first stage instrumental variables (2SLS) regression

Dependent variable dummy for innovative firms	Coef.
log employment (pre-2001)	0.084***
log age	0.[illegible]1***
(log age)^2	−0.017***
log previous experience	0.002
log(owner age)	−0.141***
Credit constrained	0.018
High-school and above	−0.006
Manufacturing	0.157***
Service	0.050*
Owner female	−0.081**
Owner has vocational training	0.185***
_cons	0.479**

Note: regional location is controlled in the estimation.
(1) owner female = 0
(2) vocational training = 0
F (2, 924) = 12.14Prob > F = 0.0000
First-stage regression summary statistics
Figures in parenthesis are standard errors, and *** $p < 0.01$, ** $p < 0.05$, * $p < 0.$

			Adjusted	Partial		
Variable		R-sq.	R-sq.	R-sq.	F(2,950)	Prob
Dummy	for	0.1492	0.1344	0.0256	12.138	0

Appendix table 6.2 Summary of the estimation results of innovation and growth equations

Variables	Innovation	Growth
Firm growth	Insignificant	Not included
Innovative firm dummy	Not included	+
log(initial employment)	+	–
log(firm age)	+	–
log(firm age)^2	–	+
log(previous experience)	Insignificant	+
log(owner age)	–	+
Female-owned	–	Not included
Credit constrained	Insignificant	–
High-school and above	Insignificant	+
Vocational training	+	Not included
Manufacturing [a]	+	+
Service [a]	Insignificant	+

Note: [a] denotes the variables are not always statistically significant.

Notes

1. CSA defines urban informal activity as those unincorporated enterprises with fewer than ten employees, no book accounts, and no licence—basically microenterprises. Enterprises with less than ten workers are also customarily classified as microenterprises in other countries, for example the European Community defines micro as firms that have zero to nine workers and small firms with ten to 99 workers.
2. The sampling frame was stratified by location and sector. Based on the population of microenterprises six major cities were first chosen, then the sample was distributed to the cities. Similar stratification across sectors was also made based on the intensity of the sector activities such as manufacturing, service, and trade. At last, a sample was taken randomly from each sector at each location.
3. In our calculation of employment we did not include casual workers, as about 80 per cent of the establishments reported that they do not normally hire casual workers.
4. It was found that in the African case studies, everything the researcher did not expect, given the traditional context and way of doing things can be called an innovation in the local context. This means making a different product or a product of slightly better quality. 'Innovation' would include all the following: using different raw materials or economizing on the use of raw materials or energy, improving the design or introducing a new way to finance, distribute, or stock products, and changing the management of a small business.
5. The debate mainly surrounds whether small or large firms are more innovative. Some argue innovations are primarily produced by large firms and concentrated markets, while others claim that small firms are more likely to innovate. The advantage of large firms on innovation is their deeper level of specialization, science-based knowledge, economy of scale, larger and cheaper financial resources, spread of risks. The strength of small firms on the other hand, lies in their flexibility, greater motivation, tacit knowledge in unique skills, more informal communication along

shorter lines, less bureaucracy, greater proximity to market, and to own production (Nooteboom 1994).

6. For more discussion and examples on this, see StataCorp Release 10, Reference I-P: 49.
7. The test of overidentifying restriction tests two things simultaneously: whether the instruments are correlated with the error term and if the equation is misspecified, i.e. one or more of the excluded exogenous variables should in fact be included in the structural equation. Thus, a significant test statistic could represent either an invalid instrument or incorrectly specified equation (StataCorp Release 10, Reference I-P: 52).
8. We have also used other estimation methods (generalized method of moments and limited information maximum likelihood) and found almost identical results, thus not reported here for brevity.
9. Take an example of two firms X and Y that had one to four employees at the initial period. If both firms add one more worker then the simple growth calculation will give 100 per cent growth for firm X and only 25 per cent growth for firm Y. The bias remains even if we take the average of current and initial size in the denominator, which gives 66 per cent and 40 per cent respectively.

References

Biggs, T., and P. Srivastava (1996) 'Structural Aspects of Manufacturing in Sub-Saharan Africa: Findings from a Seven Country Enterprise Survey', Discussion Paper 346. Washington, DC: World Bank, Africa Technical Department Series.

Bigsten A., P. Collier, S. Dercon, B. Gauhier, M. Fafchamps, J. W. Gunning, A. Oduro, R. Oostendorp, C. Pattillo, M. Söderbom, F. Teal, and A. Zeufack (2003) 'Credit Constraint in Manufacturing Enterprises in Africa', *Journal of African Economies*, 12 (1): 104–25.

Bigsten, A., and M. Gebreeyesus (2007) 'The Small, the Young, and the Productive: Determinants of Manufacturing Firm Growth in Ethiopia'. *Economic Development and Cultural Change*, 55: 813–40.

Blaug, M. (1999) 'The Concept of Entrepreneurship in the History of Economics'. In M. Blaug, *Not only an Economist: Recent Essays*. Cheltenham: Edward Elgar.

Brush, C. (1992) 'Research on Women Business Owners: Past Trends, a New Perspective and Future Directions', *Entrepreneurship Theory and Practice*, 16 (4): 5–30.

Cabral, L., and J. Mata (2003) 'On the Evolution of the Firm Size Distribution: Facts and Theory', *American Economic Review*, 93 (4): 1075–90.

Central Statistical Agency (CSA) (various years) Report on Urban Informal Sector Sample Survey, various issues. Addis Ababa: CSA.

Chipika, S., and G. Wilson (2006) 'Enabling Technological Learning among Light Engineering SMEs in Zimbabwe through Networking', *Technovation*, 26 (8): 969–79.

Cho, H.-J., and V. Pucik (2005) 'Relationship between Innovativeness, Quality, Growth, Profitability, and Market Values', *Strategic Management Journal*, 26 (6): 555–75.

Davidson P. (1991) 'Continued, Entrepreneurship: Ability, Need and Opportunity as Determinants of Small Firm Growth', *Journal of Business Venturing*, 6: 405–29.

Davidson, P., and B. Honig (2003) 'The Role of Social and Human Capital among Nascent Entrepreneurs', *Journal of Business Venturing*, 18: 301–31.

Deraniyagala, S., and H. Semboja (1999) 'Trade Liberalization, Firm Performance and Technology Upgrading in Tanzania'. In S. Lall (ed.), *The Technological Response to Import Liberalization in Sub-Saharan Africa*. London: Macmillan.

Donckels, R., and E. Fröhlich (1991) 'Are Family Businesses Really Different? European Experience from STRATOS', *Family Business Review*, 4 (2): 149–61.

Dunne, P., and A. Hughes (1994) 'Age, Size, Growth and Survival: UK Companies in the 1980s', *Journal of Industrial Economics*, 42: 115–40.

Dyer, J. W., and W. Handler (1994) 'Entrepreneurship and Family Business: Exploring the Connections', *Entrepreneurship Theory and Practice*, 19 (1): 71–84.

Elston, J. (2002) 'An Examination of the Relationship between Firm Size, Growth, and Liquidity in the Neuer Market', Discussion Paper 15/02. Frankfurt: Deutsche Bundesbank, Research Center.

Ericson, R., and A. Pakes (1995) 'Markov-perfect Industry Dynamics: A Framework for Empirical Works', *Review of Economic Studies*, 62 (1): 53–82.

Ernst, U. F. W. (2004) 'Hidden Sources of Growth? Looking at Microenterprises through the Competitiveness Lens: A Review of Evidence', Microreport 7. Washington, DC: USAID.

Evans, D. E. (1987) 'The Relationship between Firm Growth, Size and Age: Estimates for 100 Manufacturing Industries', *The Journal of Industrial Economics*, 35: 567–82.

Fagerberg, J. (2005) 'Innovation: A Guide to the Literature'. In J. Fagerberg, D. C. Mowery, and R. R. Nelson (eds), *The Oxford Handbook of Innovation*. Oxford: Oxford University Press.

Garson, G. D. (1998/2008) 'Statnotes: Topics in Multivariate Analysis'. Available at: <http://www2.chass.ncsu.edu/garson/pa765/discrim.htm>.

Geroski, P. A. (1995) 'Entry, Innovation and Productive Growth', *The Review of Economics and Statistics*, 71 (4): 572–8.

Hall, B. H. (1987) 'The Relationship between Firm Size and Firm Growth in the US Manufacturing Sector', *Journal of Industrial Economics*, 36: 583–606.

Hausman, A. (2005) 'Innovativeness Among Small Businesses: Theory and Propositions for Future Research', *Industrial Marketing Management*, 34 (8): 773–82.

Holtz-Eakin, D., D. Joulfaian, and H. Rosen (1994) 'Sticking it out: Entrepreneurial Survival and Liquidity Constraints', *Journal of Political Economy*, 102 (1): 53–75.

Hyvärinen, L. (1990) 'Innovativeness and its Indicators in Small and Medium-sized Industrial Enterprises', *International Small Business Journal*, 9 (1): 64–79.

International Labour Office (ILO) (2002) 'Women and Men in the Informal Economy: A Statistical Picture'. Geneva: ILO.

Jones-Evans, D., P. Heydebreck, M. Lindqvist, E. Autio, and M. Fontes (1996) 'Technology, Innovation and Enterprise—the European Experience'. Paper presented at the workshop on the ICSB 41st World Conference, Stockholm, Sweden, 17–19 June.

Jovanovic, B. (1982) 'Selection and the Evolution of Industry', *Econometrica*, 50: 649–70.

Khan, A. M., and V. Manopichetwattana (1989) 'Innovative and Non-innovative Small Firms: Types and Characteristics', *Management Science*, 35 (5): 597–606.

Kiggundu, M. N. (2002) 'Entrepreneurs and Entrepreneurship in Africa: What is Known and What Needs to be Done', *Journal of Developmental Entrepreneurship*, 7 (3): 239–58.

Kim, L., and Y. Kim (1985) 'Innovation in a Newly Industrializing Country: A Multiple Discriminant Analysis', *Management Science*, 31 (3): 312–22.

Koellinger, P. (2008) 'The Relationship between Technology, Innovation, and Firm Performance—Empirical Evidence from E-business in Europe', Working Paper ERS-2008-031-ORG. Rotterdam: ERIM.

Liedholm, C., and D. Mead (1993) 'The Structure and Growth of Microenterprises in Southern and Eastern Africa: Evidence from Recent Surveys', Working Paper 36. Washington, DC: GEMINI.

Mahemba, C. M., and E. J. de Bruijn (2003) 'Innovation Activities by Small and Medium-sized Manufacturing Enterprises in Tanzania', *Creativity and Innovation Management*, 12 (3): 162–73.

Mansfield, E. (1962) 'Entry, Innovation, and the Growth of Firms', *American Economic Review*, 52: 1023–51.

McDaniel, B. A. (2000) 'A Survey on Entrepreneurship and Innovation', *The Social Science Journal*, 37 (2): 277–84.

McPherson, M. (1996) 'Growth of Micro and Small Enterprises in Southern Africa', *Journal of Development Economics*, 48: 253–77.

Mead, D. and C. Liedholm (1998) 'The Dynamics of Micro and Small Enterprises in Developing Countries', *World Development*, 26 (1): 61–74.

Minitti, M., W. D. Bygrave, and E. Autio (2006) *Global Entrepreneurship Monitor: 2005 Executive Report*. Babson Park, MA: Babson College and London: London Business School.

Moreno, A. M., and J. C. Casillas (2007) 'High-Growth SMEs versus Non-high-Growth SMEs: A Discriminant Analysis', *Entrepreneurship & Regional Development*, 19 (January): 69–88.

Nooteboom, B. (1994) 'Innovation and Diffusion in Small Firms: Theory and Evidence', *Small Business Economics*, 6 (5): 327–47.

Ostlund, L. E. (1974). 'Perceived Innovation Attributes as Predictors of Innovativeness', *Journal of Consumer Research*, 1 (2): 23–9.

Oyelaran-Oyeyinka, B. (2006) *Learning to Compete in African Industry: Institutions and Technology in Development*. Farnham: Ashgate Publishing.

Penrose, E. (1959) *The Theory of the Growth of the Firm*. New York: Wiley.

Pohar, M., M. Blas, and S. Turk (2004) 'Comparison of Logistic Regression and Linear Discriminant Analysis: A Simulation Study', *Metodoloski Zverzki*, 1 (1): 143–61.

Pyke, F. (1994) 'Small Firms, Technical Services and Inter-firm Cooperation', Research Series 99. Geneva: International Institute for Labour Studies.

Robson, P. J. A., M. Haugh, and B. Obeng (2008) 'Entrepreneurship and Innovation in Ghana: Enterprising Africa', *Small Business Economics*, 32 (3): 331–50.

Rogers, E. (1995) *Diffusion of Innovations*. New York: The Free Press.

Romijn, H. (2002) 'Small Enterprise Development in Developing Countries: Innovation or Acquisition of Technological Capability?'. In M. P. van Dijk and H. Sandee (eds), *Innovation and Small Enterprises in the Third World*. Cheltenham: Edward Elgar.

Rutashobya, L. (2001) 'Female Entrepreneurship in Tanzania: Constraints and Strategic Considerations'. Proceedings of the International Academy of African Business and Development, Washington, DC, 4–7 April.

Schumpeter, J. (1939) *Business Cycles—A Theoretical, Historical and Statistical Analysis of the Capitalist Process*. New York, Toronto, London: McGraw-Hill Book Company.

Statacorp (2007) 'Stata Statistical Software: Release 10'. College Station, TX.

Stock, J., J. Wright, and M. Yogo (2002) 'A Survey of Weak Instruments and Weak Identification in Generalized Method of Moments', *Journal of Business and Economic Statistics*, 20: 518–29.

Sverrisson, A. (1997). 'Enterprise Networks and Technological Change: Aspects of Light Engineering and Metal Working in Accra'. In M. P. van Dijk and R. Rabellotti (eds), *Enterprise Clusters and Networks in Developing Countries*. Milton Park: Routledge.

Van Dijk, M. P. (2002) 'Innovation and Small Enterprise Development Examples from Burkina Faso, Ghana and Zimbabwe'. In M. P. Van Dijk and H. Sandee (eds), *Innovation and Small Enterprises in the Third World*. Cheltenham: Edward Elgar.

Van Dijk, M. P., and H. Sandee (2002) 'Innovation and Small Enterprise Development in Developing Countries'. In M. P. Van Dijk and H. Sandee (eds), *Innovation and Small Enterprises in the Third World*. Cheltenham: Edward Elgar.

Wignaraja, G. (2002) 'Firm Size, Technological Capabilities and Market-oriented Policies in Mauritius', *Oxford Development Studies*, 30 (1): 87–105.

7

Entrepreneurship and the National System of Innovation

What is Missing in Turkey?

Elif Bascavusoglu-Moreau

7.1 Introduction

Small and medium-sized enterprises (SMEs) have emerged as important agents of industrial growth since the 1980s. It is now generally acknowledged that SMEs increase overall efficiency: they are considered to be the key to the development of technology and to the knowledge-driven economy, bringing innovation to the market. In this context, entrepreneurship appears to be the mechanism that converts knowledge into growth (Acs et al. 2004).[1]

The positive impact of entrepreneurship arises through a number of mechanisms, such as knowledge spillovers (Romer 1986; Lucas 1988; Grossman and Helpman 1991), enhanced competition (Porter 1990; Glaeser et al. 1992; Feldman and Audretsch 1999), or increased diversity. But for these mechanisms to work properly, a set of well connected and interacting institutions is needed. All processes of generation, accumulation, adoption, as well as the imitation of new knowledge involve different actors such as firms, the government, research institutions, or labour and trade unions. This brings us to the 'National System of Innovation' (NIS) framework, initiated by Freeman (1987). Edquist and Johnson (1997: 14) define a system of innovation as 'all important economic, social, political, organizational, and other factors that influence the development, diffusion, and use of innovations'.

Even though the notions of NIS and entrepreneurship seem to be interdependent and complementary, the studies that combine these two approaches are rather scarce. There have been very few attempts to reconcile the notions of a NIS and entrepreneurship. By studying the case of Ireland, the Netherlands, and Finland, Golden, Higgins, and Lee (2003) assess the impact of the existence of NIS on entrepreneurship, but fail to find any correlation between institutional indicators and entrepreneurship. Radosevic (2007) presents an exhaustive and critical survey of literature on both NIS and entrepreneurship, and proposes a common basis for integrating both notions. In his framework, entrepreneurship is a systemic phenomenon driven by complementarities between technological, market, and institutional opportunities, which are matched through the NIS (Radosevic 2007).

This chapter uses this framework to analyse the impact of entrepreneurship and the NIS on the innovative capabilities of Turkish small and medium-sized firms. Innovation capabilities are defined as the skills and knowledge required to make independent adaptations and improvements to existing technologies, and ultimately to create entirely new technologies (Romijn and Albaladejo 2002), and they are measured here by innovative outputs.[2] Although very dynamic and flexible, Turkish SMEs are less innovative than their European counterparts; only 31 per cent of Turkish small firms have reported innovative activities between 2002 and 2004, compared to 42 per cent for the Europe-27, and 34 per cent for the new members (Eurostat, CIS4, and the Turkish Statistical Institute Innovation Survey). The analysis undertaken in this chapter allows us to assess whether this low level of innovative activities is related to a lack of entrepreneurial behaviour and/or to the weaknesses of the Turkish innovation system. In this sense, exploring the determinants of innovative capabilities leads to interesting discussions about the policy issues.

We explore a survey realized among 50,000 SMEs, by the Turkish Small and Medium Sized Industry Development Organization (KOSGEB). Our objective is to evaluate the impact of firms' behaviour and choices in the presence of entrepreneurial, technological, institutional, and market opportunities, in an emerging country context.

Following Radosevic, we argue that the outcome of successful entrepreneurship, here measured by innovative capabilities, is an evolutionary process of realization and the interaction of different opportunities. The first of these opportunities is technological: it is not possible to fully assess the knowledge creation process without taking technological opportunities into account (Griliches 1991; Malerba and Orsenigo 1997). Technological opportunities can be defined as the probability of innovation and/or technological improvement, with a focus on the science- and technology level within and between the sectors (Dosi 1993; Nelson and Winter 1982). The second one relates to the market: although the NIS involves both market and non-market interactions, market opportunities refer here to market-led innovation, in the lines of Kirzner (1973). Entrepreneurs are motivated by the anticipation of profits, and therefore have to be 'alert' to search the market for opportunities and innovations that bring improved goods to the market place. The third opportunity concerns institutions, which received increasing attention in shaping and driving social change and economic performance. Features identified as creating increasing returns for technologies can also be applied to institutions, representing the institutional opportunities (North 1990). Finally, entrepreneurship is approximated by both entrepreneurial behaviour and entrepreneurship capital.[3] While start-ups are used to measure entrepreneurial behaviour (Carree and Thurik 2003), the entrepreneurship index (Pınarcıoğlu and Işık 2004) reflects the entrepreneurship capital at the district (NUTS 3) level. This approximation of entrepreneurship at two levels allows for simultaneous consideration of firms' behaviour and their environment.

7.2 Small and medium-sized enterprises, entrepreneurship, and the national system of innovation in Turkey

According to the latest census of industry and business establishments, the Turkish firm's average size is 3.68 persons and enterprises employing one to 49 persons constitute 99.41 per cent of the total enterprises in Turkey (TSI 2002). SMEs contribute significantly to the Turkish economy in terms of employment (61.1 per cent), but less in terms of value added (27.3 per cent) (OECD 2002). Growth rates of the last decade relied heavily on the SMEs, whose dynamism comes from a high level of profitability, and a highly flexible labour market (Sgard, Unal, and Zlotowski 2004). However, compared to other OECD countries, their share in investments, innovation, and exports remains rather low (OECD 2005). As an example, the share of Turkish SMEs in total exports is around 9 per cent, whilst it is around 16 per cent in EU, −19.30 per cent in Hungary, 20 per cent in South Korea, and 40 per cent in India.

This weak performance of Turkish SMEs is largely related to the overall economic situation. For a long time, Turkey's economy has been characterized by high inflation, high real interest rates, and public sector imbalances, leading to repetitive crises, the last one in 2001. Although the new stability programme has resulted in a recovery, with a growth rate of 7.5 per cent between 2002–6, Turkey is only at the 75th position according to the international GDP per capita ranking. Turkey failed to implement adequate productive and technological policies to accompany its export promotion adopted in the early 1980s: it is far below the EU average in imports, exports, and especially, high-tech exports.

One of the main problems of the Turkish economy is the shadow economy. Estimated at around 30 per cent of the GDP between 1990–2003 (Schneider 2005), it is a major obstacle to productivity, competition, and innovation. This high level of unregistered businesses creates imbalances between the formal and informal sector, and somehow contributes to the high flexibility of the Turkish small and medium-sized industry (Taymaz and Ozler 2004). Although informality reduces firms' costs and provides them with the flexibility to survive under difficult conditions, it also limits their access to capital markets, their investment capacity, and their ability to develop international partnerships, restricting the potential efficiency gains that they could achieve, and trapping potential entrepreneurs in low productivity sectors (OECD 2003).

The high number of unregistered businesses can also be explained partly by the current tax legislation, which imposes a heavy financial burden on Turkish SMEs. Turkey does not have a tax integration structure to avoid the double taxation of income earned from equity investment: with a corporate tax rate of 33 per cent and a personal tax rate of 45 per cent, the total tax rate can yield to 63 per cent on fully distributed income from equity investments (OECD 2005). In this context, some entrepreneurs choose to stay in the informal sector in order to avoid excessive taxation, and hence miss the financial and technological opportunities offered to formal SMEs.

The one indicator where Turkey performs above the average for new EU entrants and candidates is the time required to start a new business (Table 7.1). Starting a business seems to take less time and require less capital than the regional average (World Bank 2004). This favorable situation is further improved by the recent laws on administrative requirements for starting companies. In 2005, the year of our empirical analysis, 3676 new firms and 3230 trade names have been established, with an increase of respectively 11.9 per cent and 18.7 per cent compared to 2004 (TSI 2006), and Turkish entrants are reported to be 40 per cent smaller than their incumbents (Taymaz 2005). This last evidence strengthens our hypothesis of using start-ups as approximation of entrepreneurial behaviour.

As put by the global competitiveness report, market efficiency has also benefited from the recent reforms, which aimed to reduce bureaucracy (WEF 2005). Turkey has improved its performance, moving from the 71st rank to the 53rd in 2005 and registered the highest rate of dynamism in upgrading competitiveness (WEF 2008). It is above the EU new member average concerning the business sophistication sub-index, particularly for the quality and quantity of networks and supporting industries. It has sophisticated industrial and service sectors which are already operating at high levels of efficiency, adopting advanced technologies, efficient production processes, and exploiting economies of scale compared to the EU new members in Central and Eastern Europe (WEF 2005). However, there is need for improvement especially in some very basic points: Turkey performs very poorly in macroeconomic and educational issues and in infrastructure quality (see Table 7.1).

Table 7.1 Comparative indicators between Turkey and new EU member/candidate countries

	Estonia	Czech Rep.	Slovenia	Slovak Rep.	Turkey	Poland	Romania	Bulgaria	Croatia	Hungary	Latvia	Lithuania	Average
GDP growth**	10.47	6.08	4.03	6.04	7.38	3.55	4.1	6.25	4.28	4.25	10.6	7.56	6.21
GNI per capita	9530	11150	17430	8100	4750	7150	3830	3510	8350	10210	6760	6910	8140
Population growth**	−0.21	0.27	0.18	0.08	1.29	−0.04	−0.23	−0.53	0.01	−0.2	−0.53	−0.62	0.1
Gross capital formation*	35.25	26.06	26.05	29.19	24.77	19.24	22.63	27.44	30.98	23.63	34.39	25.06	27.06
Industry VA*	28.51	38.29	34.08	31.66	23.69	30.79	35	30.89	30.41	30.09	21.58	33.87	30.7
Agriculture VA*	3.73	2.9	2.53	3.85	11.85	4.64	10.14	9.41	6.9	4.32	3.97	5.6	5.82
Services VA*	67.75	58.81	63.4	64.49	64.45	64.57	54.86	59.7	62.69	65.58	74.46	60.53	63.44
Exports of G\S*	79.95	71.73	64.59	77.28	27.42	37.18	32.95	59.6	46.61	66.4	47.85	58.26	55.82
Imports of G\S*	86.08	69.82	65.14	82.35	33.98	37.53	43.31	75.82	55.21	67.79	62.23	65.3	62.05
High tech exports	17.57	–	4.61	7.28	1.51	3.85	3.43	4.68	11.51	24.54	5.29	6.12	11.39
Merchandise trade*	130.23	124.71	115.23	142.24	52.36	63.03	68.99	110	70.27	117.33	86.39	106.64	98.95
Inflation**	6.79	0.68	1.47	2.37	5.39	2.58	12.29	3.76	3.19	2.01	10.18	5.77	4.71
Market capitalization*	25.42	30.93	22.99	9.26	44.46	30.96	20.83	18.71	33.21	29.52	15.75	31.9	26.16
Military expenditure*	1.48	1.78	1.67	1.74	3.23	1.79	2.11	2.35	1.58	1.31	1.69	1.8	1.88
Time required to start a business (days)	35	40	60	25	9	31	11	32	49	38	16	26	31
Tertiary school enrolment	65.74	47.99	81.19	40.63	31.19	63.41	44.79	43.64	–	65.22	74.36	76.04	55.62
Global CI score	5.12	4.74	4.64	4.55	4.14	4.3	4.02	3.95	4.26	4.52	4.57	4.54	4.59
Institutions score	4.7	3.8	4.3	4	4.05	3.6	3.4	3.1	3.7	4.2	4.1	3.9	4.17
Infrastructure score	4.66	4.5	4.51	4.08	3.46	3.64	3.05	3.41	3.98	4.05	4.33	4.14	4.28
Macro-economy score	5.31	4.81	5.08	4.37	3.58	4.34	3.91	4.92	4.3	3.94	4.93	4.82	4.62

continued

Table 7.1 continued

	Estonia	Czech Rep.	Slovenia	Slovak Rep.	Turkey	Poland	Romania	Bulgaria	Croatia	Hungary	Latvia	Lithuania	Average
Health/primary education score	6.58	6.42	6.83	6.31	6.28	6.76	6.38	6.61	6.38	6.39	6.27	37	6.54
Higher education training score	5.26	5.04	5.07	4.52	4.15	4.79	4.34	4.05	4.43	4.93	5.01	4.97	4.84
Market efficiency score	4.98	4.43	4.17	4.66	4.35	4.16	4.03	3.75	4.11	4.61	4.44	4.35	4.4
Technological readiness score	5.29	4.74	4.51	4.5	3.5	3.56	3.59	3.21	3.68	4.17	3.98	3.99	4.38
Business sophistication score	4.65	4.96	4.64	4.41	4.58	4.13	3.89	3.59	4.17	4.34	4.28	4.56	4.46
Innovation score	3.83	3.98	3.71	3.51	3.35	3.47	3.14	2.93	3.45	3.82	3.19	3.35	3.54

Notes: * % of GDP; ** annual %, *** % of manufactured exports, **** 2000 =100. Cyprus has been removed because of the scarcity of its data.

Source: All the scores are from the Global Competitiveness Index (World Economic Forum). The rest of the indicators are from World Development Indicators (World Bank).

The total entrepreneurial activity in Turkey is below the overall average (WEF 2007). Education on entrepreneurship is a very recent phenomenon, still quite limited, and the number of consultancy firms serving entrepreneurs is inadequate (TÜSİAD 2003). However, a positive attitude towards entrepreneurship has also been highlighted. Seventy-seven per cent of the population considers entrepreneurship to be a desirable choice and 86 per cent value successful entrepreneurs (Harding and Bosma 2007). Even though being risk-averse in their employment choice, Turkish people show better than average performance in terms of their approach to hard work and innovation, but without expecting a long-term positive economic effect from basic research. According to a survey realized by the Turkish Industrialists and Businessmen's Association (TÜSİAD), in starting business, Turkish entrepreneurs are mainly motivated by independence (47 per cent), creating employment (38 per cent), and earning more money (34 per cent) (TÜSİAD 2002). Although 56 per cent of the firms participating in this survey were producing new technology, only 23 per cent have been granted a patent. Twenty per cent of the firms consider the lack of innovation as a risk in business.[4]

The survey's findings match perfectly with the latest innovation survey conducted by the National Institute of Statistics. This latter shows that 22 per cent of enterprises with ten or more employees are product (goods and services) innovators and 22.6 per cent process innovators: while 31.4 per cent of enterprises in Turkey were active in developing or implementing innovations for the period 2004–6. Innovation frequency increases with firm size—with large enterprises with more than 250 employees being more likely to engage in some sort of innovation activity (43.5 per cent) than smaller enterprises (29.7 per cent).

Given the high rates of inflation in the past, combined with the high dependency on foreign technology, the rise of innovation costs appear to be the main obstacle for Turkish innovators: 69.2 per cent of the enterprises report high innovation costs as the main obstacle for innovation, followed by lack of qualified personnel (65.7 per cent), and lack of internal or external funding (65.3 per cent). This is particularly relevant for SMEs, which are not able to reflect the increasing input prices in the selling prices, see their already insufficient equities diminish, leaving them with increasing financial needs. In line with European harmonization, Turkey has recently adopted policy changes in the finance, technology, and competitiveness areas, where improving SMEs and entrepreneurship stands out as one of the main objectives.[5] Investments in innovation are promoted through tax incentives, matching grants, and reimbursable loan schemes. There are four main fiscal provisions to support R&D in Turkey, although fiscal incentives often do not benefit SMEs, which have insufficient profits to use the tax benefits and do not record R&D expenses separately on financial statements (World Bank 2004).

The last European trend chart report reckons that Turkey has almost every element that makes up a NIS: a broad policy mix, with a wide range of instruments and measures in almost all areas of innovation policy, as laid down in the European Union Action Plan of 1996 (EC 2005). The major weakness of the Turkish innovation system lies in the lack of cooperation and linkages between different actors. The number of research collaborations between university and industry are relatively low compared to most EU and Asian countries (WEF 2007). It has been reported that 'most of the firms in Turkish manufacturing industries do not work with any research center or university in Turkey or abroad in acquiring knowledge or in developing new technologies' (Taymaz 2005: 12), and the main reason for this lack of cooperation is the lack of information. Furthermore, potential intermediaries between research institutions and industry are rather scarce. Even though there has been a recent increase in the number of incubators, technology parks, and technology transfer offices, especially after the 'Law on Technology Development Zones' in 2001, the number of intermediaries remains unsatisfactory given the size of the country. Moreover, the industrial and social heterogeneity of Turkish regions suggests a more decentralized approach of the NIS. In this context, the recently established regional development agencies become a crucial part of the NIS, but it is still too early to evaluate their impact.

Overall, the NIS, as well as entrepreneurship capital, present a mixed picture in Turkey. Despite some recent policy reforms, there is still room for improvement in many areas. Turkish firms, particularly small and medium-sized firms, lag behind their European counterparts. This chapter attempts to explore this poor performance of Turkish SMEs from two different perspectives: entrepreneurship and NIS.

7.3 Database, variables, and methodology

We use a unique firm level survey data collected by KOSGEB in 2005.[6] Our database covers 50,436 SMEs, and gives information about the firm characteristics such as size, age, educational level, as well as its productive, exporting, and innovative behaviour.

After removing outliers that could bias the estimation results, and all the observations with missing variables, 46.54 per cent of the remaining firms are micro-sized (one to nine employees), and only 9.5 per cent of the firms in the whole database employs more than 150 people. Although firms employing up to 250 persons are considered as SMEs in both international and Turkish classifications, we remove firms employing more than 150 persons, due to their small number in our sample and the possible bias they could introduce.

Almost 88 per cent of the firms have been founded after 1980, the same period when a major liberalization process has been adopted in Turkey.

Of the Turkish SMEs, 53.8 per cent performs in low-tech industries, followed by 26 per cent in middle low-tech, and 19 per cent in middle high-tech industries.[7] Regardless of their technological intensity, all sectors are included in the sample, as this chapter does not aim to assess knowledge-based entrepreneurship. We believe that innovation can happen at every level, particularly in the context of an emerging country. Given Turkey's overall economic situation and the SME characteristics briefly discussed above, it is likely to have improvements and/or innovations with small technological content.

This last consideration is reflected in the choice of the dependent variable. We approximate the innovative capability by an output measure; a dummy variable which equals 1 if the firm holds a patent, utility model, or trademark. Following this broad definition, 37.4 per cent of our sample have a certain level of innovative activity, whilst only 8 per cent hold a patent. Unlike patents, utility models are granted without prior research to establish novelty and inventive step. Therefore, they are cheaper and easier to obtain and present an alternative to patenting for smaller firms. Of our sample, 9 per cent has a utility model, and less than 4 per cent of the sample with innovative capabilities holds both a utility model and a patent. Although noisy, patents and utility models have been widely used as an innovation indicator in the previous literature. However, the choice of including trademarks in our innovative capability indicator might need further explanation. Trademarks are the outcome of establishing recognizable designation as well as firms' identities (Mendonca, Pereira, and Godinho 2004) and have been used as a complementary indicator in empirical innovation studies (Greenhalgh and Rogers 2007; Mendonca, Pereira, and Godinho 2004). Innovative firms are found to use the trademarks more than less innovative ones, and trademarks are used more than patents among the innovative firms (Schmoch 2003; Mendonca, Pereira, and Godinho 2004). Moreover, firms in developing countries tend to invest more in brands than in new technology (Baroncelli, Kink, and Javorcik 2004), as registration fees for trademarks are much lower than patents (Malmberg 2005). Finally, we argue that the use of a composite indicator that takes into account different aspects of innovation is more accurate, particularly in an emerging country context.

The independent variables are firm age and size, the average educational level of the employees, R&D, technology outsourcing, the use of technology-intensive equipment in the production process,[8] the use of information and communication technologies (ICTs), the possession of quality labels, and export behaviour. We also control for the firms' capital level: although we do not know the firms' net current assets, we have the information about their relative size. Our sample is categorized into four classes according to net current assets. In order to test the impact of this variable, we introduce dummies in our estimation, taking the second asset category as the reference category.[9]

7.3.1 *Entrepreneurship*

We aim to take into account simultaneously the entrepreneurial behaviour and the entrepreneurship capital, so entrepreneurship is measured at both firm- and region level.

Entrepreneurship is not synonymous with small business, but small firms represent an outstanding vehicle for individuals to channel their entrepreneurial ambitions (Carree and Thurik 2003). Although individuals in large firms, i.e. corporate entrepreneurs, can also undertake entrepreneurial actions, we are interested in persons starting or operating in innovative small firms. Studies evaluating the impact of entrepreneurship on economic performance on the level of firm or establishment use self-employment rates or the firm characteristics such as firm size and/or age. As our sample is solely composed from small and medium-scale enterprises, we choose to approximate entrepreneurial behaviour by start-ups. This variable has been widely used in the previous literature to measure entrepreneurship (Audretsch and Thurik 2001; Blanchflower 2000, Carree et al. 2000; Fonseca, Lopez-Garcia, and Pissarides 2001). We consider that a firm is 'entrepreneurial' if the firm age is equal to or less than five years, and we only consider new firms, hence family firms or purchased firms have not been taken into account.

To test the robustness of the entrepreneurship variable, we refer to the KOSGEB survey, where managers/entrepreneurs have been asked to evaluate various aspects the performance of the firm, according to a three point scale (increasing/stable/decreasing). Appendix table 7.2 gives the answers to these questions, both for the whole sample and the start-ups. It shows that start-ups perform better than the whole sample in almost every performance measure, as the increase in their performance is greater than the whole sample.[10] We can hence conclude that start-ups provide an adequate measure for entrepreneurship in Turkey.

We also take into account entrepreneurs' educational level: 40.8 per cent of the SMEs in our database have a manager with at least a university degree. This ratio is around 19 per cent for the start-ups. The average enrolment year of the overall employees is 10.5 years, which indicates a rather high level of education in the Turkish SMEs.

Finally, in order to measure the entrepreneurship capital, we introduce an exhaustive entrepreneurship index at district (NUTS 3) level. Developed by Dr Melih Pınarcıoğlu, this index accounts for districts' entrepreneurial capacity by taking into account 15 different indicators, such as export growth rate and its variety, innovativeness, firm turnover, and subventions, among others.[11] The index varies between zero and one, the lowest score being 0.182 (for Bitlis) and the highest 0.607 (for Istanbul). It highlights the high level of internal disparities in Turkey, where the Eastern part accounts for 37 per cent of population and 22 per cent of GDP, whereas the Western part accounts for respectively 63 per cent and 78 per cent (Reeves 2005).

7.3.2 *National innovation system*

Technological opportunities are reckoned to be localized and geographically bounded (Fagerberg 2003). We will therefore approximate the technological opportunities at NUTS 3 level, measuring the industrial structure of the region where the firm is located, by the total capacity of power equipment and the per capita electricity consumption in the manufacturing industry. The spatial distribution of the industry in Turkey is reported to have slightly changed in recent years (Pınarcıoğlu and Işık 2004). The service sector begins to overtake the industrial sector in traditionally industrial regional centres, and decentralized industry is beginning to concentrate in the rural provinces that are neighbours to traditional regional centres (Ozaslan, Dincer, and Ozgur 2006). Furthermore, some provinces located in different regions of Anatolia appear now as new industrial districts by specializing in certain sectors exploiting local endogenous resources. Our two variables aim to take these trends into account. We also introduce a firm level clustering variable which equals to one if the firm is localized in a particular cluster, i.e. in a small or organized industrial zone (OIZ). Of the whole sample, 22 per cent is located in a small industrial estate and 20 per cent in an OIZ. In order to account fully for the technological opportunities in a national/regional innovation system, we introduce this location variable interacted by the number of plots in industrial estates at NUTS 3 level.

Market opportunities are approximated by the location quotient and the urbanization rate imputed at a regional level. The location quotient measures the proportion of an activity in a particular area, compared to the proportion of the aggregated activities in this area at NUTS 3 level. The more concentrated the region/sector is, the more profit firms should have, hence we would expect a higher entry rate, and a greater market opportunity. The urbanization rate reflects the demographic structure of the city where the firm is located. A higher urbanization rate should reflect a greater market and therefore increase the innovative capacity.

Institutional opportunities stand for institutions, norms, and rules that affect the innovation process (Radosevic 2007). In Turkey, both small and large enterprises view high innovation costs and lack appropriate finance as main factors hampering innovation activities (Napier et al. 2004).

Furthermore, previous literature has shown that the smaller the enterprise, the bigger their financial problems. We introduce several variables which approximate the financial facilities and support available for SMEs. The first one of these variables is the consolidated budget revenue per capita at NUTS 3 level, in order to account for regional disparities in public investments. However, public investments might not be adequate for the SMEs' needs. Hence we consider investment incentives and private (bank) loans. These two variables are introduced as interactions: firms with an incentive certificate interacted with per capita amount of incentives (granted with incentive certificate)[12] and firms benefiting from a loan with the per capita amount of credits, both at NUTS 3 level.

7.3.3 *Methodology*

Our empirical specification is based on the knowledge production function (Griliches 1979), which models the 'functional relationship between the inputs of the knowledge production and its output that is economically useful new technological knowledge' (Acs et al. 2004: 1074). The basic assumption states that the output of the innovation process (Q) is a result of the resources invested in inventive activity (R), usually R&D capital or investment, and that the patents (P) are a good measure of this economically valuable knowledge. The patents do not play any explicit economic role in Griliches' model. They are just an indicator of innovative activity, based on the assumption that some random fraction of Q gets patented.

Following Griliches, we assume that Q_i is an index of innovative output of the firm i, and the stock of R&D is the main input in the innovative activity. The production of Q_i can then be expressed by a standard knowledge production function $Q_i = f(R_i, X_i, vi)$ where X represents the other variables that may affect the innovative activity, such as firm size, type of activity, and/or sector-specific effects and R stands for R&D. As mentioned in the previous section, we adopt a broad definition of knowledge production, and take into account patents, utility models, and trademarks to measure Q_i. The basic specification to be estimated in order to evaluate the determinants of Turkish SMEs' innovative capabilities is therefore:

$$q_i = \alpha_1 + \alpha_2 R_i + \alpha_3 X_i + \epsilon_i \qquad (1)$$

Following Audretsch and Keilbach (2004), we consider that entrepreneurship involves a number of different legal, institutional, and social factors: an entrepreneurship capital, measured in this chapter by the entrepreneurship index (E_j). We also introduce firms' behaviour, approximated by being a start-up (E_i) and entrepreneurs' educational level (ED_i). This second specification is given below:

$$q_i = \alpha_1 + \alpha_2 R_i + \alpha_3 X_i + \alpha_4 E_i + \alpha_5 E_j + \alpha_6 ED_i + \epsilon_i \quad (2)$$

The subscripts *i* and *j* stand respectively for firm- and regional levels. Following Radosevic (2007), the NIS is approximated by technological (*T*), institutional (*I*), and market (*M*) opportunities, as following:

$$q_i = \alpha_1 + \alpha_2 R_i + \alpha_3 X_i + \alpha_4 T_j + \alpha_5 M_j + \alpha_6 I_j + \epsilon_i \quad (3)$$

Given that the objective of the chapter is to evaluate the impact of firms' behaviour and choices in the presence of entrepreneurial, technological, institutional, and market opportunities, we will introduce several interaction terms in equations (2) and (3) between firms' choices and behaviours (F_i) and entrepreneurial, technological, and institutional opportunities, leading to final specifications:

$$q_i = \alpha_1 + \alpha_2 R_i + \alpha_3 X_i + \alpha_4 E_i + \alpha_5 E_j + \alpha_6 ED_i + \alpha_7 (E_i * E_j) + \epsilon_i \quad (2a)$$

$$\begin{aligned} q_i = \alpha_1 &+ \alpha_2 R_i + \alpha_3 X_i + \alpha_4 T_j + \alpha_5 M_j + \alpha_6 I_j + \alpha_7 (F_i * T_j) \\ &+ \alpha_8 (F_i * I_j) + \epsilon_i \end{aligned} \quad (3a)$$

As our innovative capability variable is measured as binary outcomes, the empirical model estimated is probit. Given the possibility of heteroskedasticity and clustering effects, we compute robust standard errors and introduce industry dummies at Nace 2 digit level (not reported). Although probit (and logit) models are widely used in the empirical research, there are a number of critical issues that one should be aware of when interpreting the results, given their non-linear nature.[13] In what follows, we present the variables' marginal effects, i.e. how much a change in variable changes the probability of innovation output in Turkish SMEs, by setting the other variables at their mean.

7.4 Results

7.4.1 *Determinants of Turkish small and medium-sized enterprises' innovative capabilities*

We begin our analysis by evaluating the determinants of Turkish SMEs' innovative capabilities. The various estimation results for our basic specification are presented in Table 7.2, without the entrepreneurial and NIS variables. Column

Table 7.2 Basic specifications

	(i)	(ii)	(iii)	(iv)
Firm age	0.003***	0.011***	0.011***	0.026***
	−16.03	−15.81	−14.38	−9.38
Age squared				−0.000***
				(−4.99)
Firm size	0.001***	0.002***	0.002**	0.016***
	−7.09	−6.11	−3.04	−7.46
Size squared				−0.000***
				(−8.12)
R&D	0.060***	0.195***	0.196***	0.185***
	−13.24	−13.29	−7.81	−7.15
Education	0.005***	0.019***	0.018**	0.023***
	−3.4	−3.94	−2.98	−3.58
Technology	0.018***	0.066***	0.047	0.029
	−3.77	−4.17	−1.25	−0.83
ICT	0.002***	0.007***	0.011***	0.009***
	−4.68	−3.76	−5.24	−4.95
Export	0.117***	0.354***	0.396***	0.352***
	−23.8	−22.46	−8.35	−7.23
Outsourcing	0.027***	0.082***	0.103***	0.092**
	−5.72	−5.37	−3.57	−3.28
Quality labels	0.281***	0.806***	0.784***	0.745***
	−47.64	−41.59	−11.67	−11.16
Assets 1	−0.034***	−0.130***	−0.162***	−0.134***
	(−6.60)	(−7.69)	(−5.31)	(−4.81)
Assets 3	0.044***	0.127***	0.129***	0.091*
	−6.3	−5.71	−3.85	−2.44
Assets 4	0.065***	0.187***	0.188***	0.139***
	−10.21	−9.1	−4.93	−3.5
Constant	0.408***	−0.301***	−0.497***	−0.766***
	−24.13	(−5.41)	(−5.32)	(−7.34)
Pseudo R_squared		0.19	0.179	0.187
Number of observations	41923	41923	41923	41923
Number of clusters			81	81
Chi2		8286.391	14552.977	32503.342
AIC	47167.936	45014.844	45566.098	45145.913

Note: *Significant at 10%; **significant at 5%; ***significant at 1%, except the first column, marginal effects are reported.

(1) reports the standard ordinary least squares model with fixed effects for sectors and regions (NUTS 2 level) as a benchmark. The subsequent columns report the results from the three different specifications of probit estimators: respectively a sector- and region-specific fixed effect model, a sector-specific fixed effect model with region clusters, and a sector-specific fixed effect model where the squared terms of firm age and size have been introduced.[14] This final specification (column 4) is therefore our preferred specification.[15] All the variables have the expected signs, except the use of technology, not significant in the last two columns. Investing in R&D increases the probability of innovation, as well as the exporting activity. However, it is ownership of quality labels and certificates which seems to increase the propensity to innovate the

most. These quality standards demonstrate the abilities of firms to adapt and adopt a body of specialized and codified knowledge (UNIDO 2005). Therefore they seem to be more important than R&D investments in building innovative capabilities in Turkish SMEs. We also see that the odds of innovation are larger if the firm belongs to upper classes in terms of net assets, compared to the lowest one. These results draw attention to the problem of capital access in SMEs. We also note an inverted U-shaped relationship between the firm's age and size, and the propensity to innovate. The precise estimate of the turning point after which extra size (age) affects innovative capabilities negatively is found to be 65.7 (39.9).[16]

7.4.2 *Innovative capabilities, entrepreneurship, and the national system of innovation*

In the following, we seek to analyse the impact of entrepreneurship capital and entrepreneurial behaviour on Turkish SMEs' innovative capabilities (Table 7.3). Two different specifications have been defined for that purpose. The first specification (columns 1 and 2) takes into account the firm age, and the second one (columns 3 and 4) introduces the start-up variable as defined earlier.

We note that the introduction of entrepreneurial variables does not affect significantly the rest of the parameters, except the use of technology-intensive equipment in the production process, which becomes significant once again. This result shows that the purchase of machinery and the learning of operating procedures are not sufficient for an effective technology transfer (Dahlman and Westphal 1982). It is then possible to conclude that the entrepreneur will bring the necessary skills to adopt and adapt the new technology within the firm.

When we estimate equation (2) without the interaction terms (i) and (iii), the entrepreneurial index seems to increase innovative capabilities considerably, a first evidence on the positive impact of entrepreneurship capital. When the interaction term between the firm age and entrepreneurship index is introduced (specification (ii)), our results show that the positive effect of the entrepreneurial capital on Turkish SMEs' innovative capabilities declines for older firms. Although the magnitude and statistical significance of the interaction effect varies by observation,[17] the estimated effect of the interaction term is around –0.003, with a standard error of 0.001.

The entrepreneurial behaviour as defined in the previous literature (i.e. being a start-up) has a negative sign ((iii) and (iv)). As one might have expected, not all start-ups are entrepreneurs. In fact, without considering the firm's environment, our approximation of firm level entrepreneurship risks

Table 7.3 Innovation capabilities and entrepreneurship

	(i)	(ii)	(iii)	(iv)
Firm age	0.0112***	0.0165***		
	−14.82	−5.34		
Firm size	0.0165***	0.0165***	0.0160***	0.0160***
	−8.33	−8.38	−8.03	−8.02
Size squared	−0.000123***	−0.000123***	−0.000119***	−0.000119***
	(−8.72)	(−8.73)	(−8.30)	(−8.29)
R&D	0.182***	0.181***	0.184***	0.184***
	−7.33	−7.33	−7.46	−7.44
Education	0.0178***	0.0177***	0.0182***	0.0182***
	−2.73	−2.72	−2.81	−2.81
Technology	0.0311	0.0311	0.0306	0.0305
	−0.92	−0.92	−0.9	−0.9
ICT	0.00661***	0.00668***	0.00670***	0.00672***
	−3.55	−3.57	−3.65	−3.64
Export	0.322***	0.323***	0.313***	0.313***
	−7.02	−7.03	−6.86	−6.87
Outsourcing	0.0819***	0.0815***	0.0791***	0.0791***
	−2.7	−2.68	−2.58	−2.58
Quality label	0.760***	0.760***	0.762***	0.762***
	−11.82	−11.85	−11.93	−11.93
Entrepreneur's education	0.0167	0.0171	0.0215	0.0215
	−0.81	−0.83	−1.06	−1.06
Entrepreneurship index (I_E)	0.632***	0.798***	0.632***	0.616***
	−4.07	−5.19	−4.05	−3.74
Age X I_E		−0.0123**		
		(−2.13)		
Start-up			−0.335***	−0.374***
			(−22.96)	(−6.72)
Start-up X I_E				0.0884
				−0.8
Asset 1	−0.130***	−0.130***	−0.134***	−0.134***
	(−4.26)	(−4.28)	(−4.50)	(−4.49)
Asset 3	0.0973***	0.0971***	0.0908**	0.0908**
	−2.59	−2.6	−2.48	−2.48
Asset 4	0.150***	0.150***	0.148***	0.148***
	−3.66	−3.66	−3.68	−3.68
Constant	−0.845***	−0.917***	−0.619***	−0.612***
	(−6.17)	(−6.55)	(−4.46)	(−4.27)
Pseudo R-squared	0.187	0.188	0.189	0.189
Number of observations	41923	41923	41923	41923
Number of clusters	81	81	81	81
Chi2	14901.9	26144.89	26458.31	52139.67
AIC	45112.01	45109.12	45011.61	45013.22

Notes: *Significant at 10%; **significant at 5%; ***significant at 1%, marginal effects are reported.

measuring also the individuals who started a new company not by entrepreneurial motivations but because they had no other choice. We therefore have to refine our specification and account for the complex relationship between innovation system and entrepreneurship.

For this purpose, we introduce an interaction term between entrepreneurial behaviour and entrepreneurship index, aiming to fully capture the impact of the firm's environment, estimating thus equation (2a). Despite the lack of statistical significance of the coefficient of the interactive term (iv), the full interaction effect is S-shaped and has an overall impact of 0.002.

Table 7.4 shows the results of the regression where the NIS variables are introduced (equation 3a). Although their intensity varies the sign of traditional determinants does not change significantly with the inclusion of NIS variables. Regarding technological opportunities, the industrial structure and capacity have expected positive signs. Firms' innovative capabilities increase with the industrial infrastructure. However, being located in larger industrial zones does not seem to affect the propensity to innovate. Industrial zones in Turkey seem to fail to promote innovative capabilities of younger, smaller, less technology-intensive, and/or more disadvantaged firms. This result can nonetheless be explained by a number of factors. First of all, not all industrial estates have technological facilities and innovative incentives. Even so, one would expect a positive impact of agglomeration economies and spillovers, arising from clusters. But if the spillovers do not have a technical nature, we can expect the positive effects being captured by the market opportunities. Furthermore, previous studies have found a low level of network and interactions in Turkey (Akçomak and Taymaz 2004) and suggested that spillovers are more likely to occur in a diversified industrial structure (Bascavusoglu 2008) which may partly explain these results.

Regarding the market opportunities, both the location quotient and the urbanization rate have expected positive values, although their impact is rather small. A location quotient greater than one indicates that the level of employment in that industry exceeds the local demand for the goods or services which it produces, thus indicating a full exploitation of local market opportunities. Small firms tend to operate more and more efficiently on the appropriate scale to the size of local markets. On the other hand, urbanization economies are acknowledged to be the scale effects associated with city size or density, they are external to industries but internal to geographic units (Feldman 1999). A positive impact of the urbanization rate on innovative capabilities points to Jacobian diversification externalities. We can therefore conclude that urbanization economics seem to be more important in Turkey.

The budget revenue per capita, first of a set of variables approximating the institutional opportunities, is significant and positive as expected. The incentive variable is also significant. It clearly shows that the amount of incentive per capita increases the probability of innovation. The final result regarding institutional opportunities is the non-significance of private loans, highlighting the seriousness of the financial problem for Turkish SMEs. Not only are private bank loans scarce and not easily available for the SMEs, but they also hamper innovative capabilities when finally obtained. The cost of short term financing is the highest in Turkey compared to the other European countries

Table 7.4 Innovative capabilities and the National System of Innovation dimension

	Technological opportunities	Market opportunities	Institutional opportunities
Firm age	0.025***	0.026***	0.026***
	−9.13	−9.39	−9.11
Age squared	−0.000***	−0.000***	−0.000***
	(−4.83)	(−4.96)	(−4.87)
Firm size	0.016***	0.016***	0.016***
	−7.62	−7.63	−7.88
Size squared	−0.000***	−0.000***	−0.000***
	(−7.86)	(−7.95)	(−8.41)
R&D	0.186***	0.179***	0.182***
	−8.33	−7.04	−7.73
Education	0.026***	0.022***	0.022***
	−4.64	−3.64	−3.45
Technology	0.037	0.032	0.029
	−1.26	−0.95	−0.93
ICT	0.007***	0.008***	0.009***
	−3.37	−3.46	−4.54
Export	0.322***	0.330***	0.346***
	−7.36	−7.81	−7.92
Outsourcing	0.080**	0.086**	0.089**
	−2.79	−2.84	−2.99
Quality labels	0.762***	0.750***	0.740***
	−10.72	−11.8	−11.31
Electricity per capita	0.038***		
	−4		
Power capacity	0.001*		
	−2.32		
Location	−0.114		
	(−1.82)		
OIZ size	−0.012		
	(−0.95)		
Location X OIZ size	0.01		
	−0.71		
Location quotient		0.061*	
		−1.84	
Urbanization rate		0.003*	
		−1.74	
Incentives			0.128***
			−3.45
Incentives per capita			0.004
			−0.34
Incentive X incentive per capita			−0.016
			(−1.92)
Budget revenue per capita			0.135*
			−0.66
Credit			−0.022
			(−0.69)
Credit per capita			−0.412
			(−1.71)
Credit X credit per capita			0.16
			−0.86
Constant	−0.879***	1.032***	−0.728***
	(−8.96)	(−4.99)	(−6.48)
Pseudo R-squared	0.191	0.188	0.188

Number of observations	41923	41923	41923
Number of clusters	81	81	81
Chi2	35804.676	39875.944	57236.005
AIC	44897.297	45076.261	45104.381

Notes: *Significant at 10%; ** significant at 5%; *** significant at 1%. Standard errors in brackets. Marginal effects are reported. Assets dummies (not reported) are included in the estimation.

(EC 2003), discouraging small firms from investing in knowledge creation, a highly risky, uncertain, and costly process.

7.5 Conclusion

The purpose of this chapter is to evaluate Turkish SMEs' innovative capabilities from two perspectives, namely the entrepreneurship and the NIS. As the starting point of our chapter is the low level of innovative activities in Turkish SMEs, we approximated innovative capabilities with a broadly defined innovative performance measure. Through a survey conducted on 50,000 SMEs, we took into account entrepreneurial, technological, market, and institutional opportunities, and their interaction with firms' behaviour and choices.

Our first conclusion concerns the determinants of innovative capabilities of Turkish SMEs. Overall, investment in R&D, the use of ICTs, exporting activities, outsourcing and employees' educational level are found to increase Turkish SMEs' innovative activities. But more particularly, owning quality labels and certificates has the most important impact on innovative performance, regardless of the model specification. These quality standards demonstrate the firms' ability to learn, adopt, and adapt specialized and codified knowledge. Even though certified management systems and/or quality labels require a considerable economical expense, especially for small firms, the expected benefits seem to be substantial. The result shows that alongside the usual positive effects such as increased profits and market shares, improved performance, the ability to meet client expectations, and facilities to participate in international markets, quality labels and certificates also improve the innovation abilities of small firms.

Our second conclusion points to the weaknesses of both the NIS and firm's entrepreneurial behaviour, where some interesting policy implications emerge.

The regional industrial structure, approximated by the total capacity of power equipment and per capita electricity consumption, is found to be important in building innovative capabilities. However, being located in an industrial estate does not increase the odds of innovation. Furthermore, we found no positive relationship between the size of the industrial estates and innovative performance. This finding stresses an important policy issue, considering that these OIZs are reported to have an excess capacity in Turkey (OECD 2004). Nonetheless, Turkey continues to provide extensive loans and new funds for the establishment of small industrial estates and/or OIZs. Our results suggest that these types of clustering efforts may not be fully adequate for Turkish SMEs' needs. Besides, the analysis of market opportunities highlights the existence of diversification externalities in most of the specification, confirming previous studies on urbanization and productivity in Turkey (Lall, Deichmann, and Coulibaly 2007; Bascavusoglu 2008).

This study also underlines the extent of financial problems faced by Turkish SMEs. The amount of private loans at regional level is not channelled towards innovative capabilities of Turkish SMEs located in that region, pointing to the scarcity of available private credit for small firms. Furthermore, a private loan does not seem to foster knowledge creation.

Finally, although the start-up firms, supposed to reflect the entrepreneurial Turkish SMEs, do not show a better performance, the regional entrepreneurship capital seems to contribute positively to the SMEs' innovative capabilities, especially for young firms. However, we are aware that start-ups might not fully reflect the entrepreneurial firms, especially in a developing country context such as in Turkey. Empirically defining entrepreneurial firms is a very interesting challenge that we would like to explore in future research.

Indeed, the Turkish small and medium-sized industry is far from the knowledge-based, internationally competitive, and innovative entrepreneurial small firms. Yet, SMEs are known for their dynamism particularly in finding market niches and benefiting from cheap inputs from the informal economy (EC 2006). Our result suggests that this dynamism and flexibility might be an outcome of entrepreneurial opportunities. The empirical analysis also shows the importance of institutional opportunities in comparison to other NIS issues. In the light of our empirical analysis, we argue that the weaknesses of Turkish SMEs are more policy-related, rather than resulting from a lack of entrepreneurial behaviour.

APPENDIX

Appendix table 7.1 Sectoral distribution

NACE	Whole sample		Start-ups	
	Freq.	Per cent	Freq.	Per cent
Basic metals	2,630	5.83	474	5.39
Chemicals and chemical products	1,488	3.3	281	3.2
Coke, refined petroleum product, and nuclear fuel	100	0.22	26	0.3
Electrical machinery and app. n.e.c.	1,442	3.2	219	2.49
Fabricated metal products, exp. machinery and equipment	3,576	7.93	639	7.27
Food products and beverages	5,328	11.81	938	10.67
Furniture, manufacturing n.e.c.	6,595	14.62	1,439	16.38
Machinery and equipment n.e.c.	4,314	9.55	703	8
Medical, precision and optical inst., watches and clocks	355	0.79	95	1.08
Motor vehicles, trailers, and semi-trailers	1,053	2.33	160	1.82
Office machinery and computers	74	0.16	15	0.17
Other non-metallic mineral products	2,406	5.33	464	5.28
Other transport equipments	273	0.61	42	0.48
Publishing, printing, and reproduction of recorded media	505	1.12	119	1.35
Pulp, paper, and paper products	820	1.82	150	1.71
Radio, television, and communication equipment	137	0.3	24	0.27
Recycling	143	0.32	38	0.43
Rubber and plastic products	3,017	6.69	600	6.83
Tanning and dressing of leather	1,467	3.25	275	3.13
Textiles	6,738	14.93	1,517	17.26
Tobacco products	129	0.29	31	0.35
Wearing ap., dressing and dyeing of fur	1,468	3.25	351	3.99
Wood, products of wood and cork, excl. furniture	1,062	2.35	187	2.13
Total	45,120	100	8,787	100

Source: KOSGEB.

Appendix table 7.2 Performance evaluation

Performance indicator	Whole sample		Start-ups	
	Freq.	Per cent	Freq.	Per cent
Sales				
Increase	26,034	61.9	5,128	68.74
Stable	9,280	22.06	1,581	21.19
Decrease	6,747	16.04	751	10.07
Costumer returns				
Increase	1,712	5.01	232	3.85
Stable	17,949	52.56	3,199	53.09
Decrease	14,491	42.43	2,595	43.06
Wasted products				
Increase	3,280	8.57	498	7.37
Stable	19,882	51.95	3,376	49.98
Decrease	15,107	39.48	2,881	42.65

continued

Appendix table 7.2 continued

Performance indicator	Whole sample		Start-ups	
	Freq.	Per cent	Freq.	Per cent
Machinery park				
Increase	23,657	56.83	4,278	58.32
Stable	17,044	40.94	2,949	40.2
Decrease	926	2.22	109	1.49
Customer satisfaction				
Increase	35,538	83.55	6,488	86.15
Stable	6,454	15.17	974	12.93
Decrease	544	1.28	69	0.92
Efficiency of transport				
Increase	29,619	72.21	5,300	73.27
Stable	10,662	25.99	1,816	25.1
Decrease	737	1.8	118	1.63
Product variety				
Increase	31,000	72.83	5,630	74.73
Stable	11,015	25.88	1,832	24.32
Decrease	547	1.29	72	0.96
Costumers				
Increase	31,017	73.14	5,902	78.62
Stable	8,897	20.98	1,351	18
Decrease	2,494	5.88	254	3.38
Production				
Increase	28,709	68.86	5,585	75.7
Stable	9,221	22.12	1,422	19.27
Decrease	3,764	9.03	371	5.03
Exports				
Increase	10,443	41.91	1,633	40.45
Stable	12,399	49.76	2,188	54.2
Decrease	2,075	8.33	216	5.35
Capital				
Increase	22,273	54.05	3,843	52.8
Stable	16,336	39.64	3,138	43.11
Decrease	2,598	6.3	298	4.09
Total debt				
Increase	12,192	32.13	2,075	30.66
Stable	18,338	48.33	3,150	46.55
Decrease	7,417	19.55	1,542	22.79
Productivity				
Increase	27,973	68.09	5,333	73.12
Stable	10,992	26.75	1,703	23.35
Decrease	2,119	5.16	257	3.52
Efficient use of personnel				
Increase	24,923	61.59	4,665	65.38
Stable	13,760	34	2,242	31.42
Decrease	1,786	4.41	228	3.2
Export countries				
Increase	9,392	37.56	1,438	35.39
Stable	14,228	56.91	2,460	60.55
Decrease	1,383	5.53	165	4.06
Capacity utilization rate				
Increase	23,242	57.45	4,602	64.3
Stable	13,451	33.25	2,148	30.01
Decrease	3,765	9.31	407	5.69

Cost of production				
Increase	29,529	70.55	4,681	63.32
Stable	9,166	21.9	2,022	27.35
Decrease	3,162	7.55	590	9.33
Competitiveness				
Increase	26,389	63.89	4,947	67.65
Stable	9,054	21.92	1,618	22.12
Decrease	5,864	14.2	748	10.23
Profitability				
Increase	14,167	34.87	2,927	40.79
Stable	14,082	34.66	2,779	38.73
Decrease	12,384	30.48	1,469	20.47
Distributed profit				
Increase	7,537	25	1 506	28.81
Stable	15,838	52.53	2 894	55.37
Decrease	6,778	22.48	827	15.82

Source: KOSGEB.

Appendix table 7.3 Variables, descriptions, and statistics

(a) Summary statistics

Variable	Obs.	Mean	Std. dev.
Innovation	41923	0.37	0.48
Firm age	41923	13.59	9.87
Firm size	41923	19.78	25.01
R&D	41923	0.36	0.48
Education	41923	9.84	1.53
Technology	41923	0.29	0.45
ICT	41923	3.44	5.66
Export	41923	0.37	0.48
Outsourcing	41923	0.33	0.47
Quality labels	41923	0.18	0.39
Entrepreneur's education	41923	0.37	0.48
Entrepreneurship index	41923	0.43	0.12
Start-up	41923	0.20	0.40
Electricity per capita	41923	3.33	2.57
Power capacity	41923	60.79	69.90
Location	41923	0.44	0.49
OIZ size	41923	4.25	3.11
Location quotient	41923	1.16	0.48
Urbanization rate	41923	72.80	16.24
Incentives	41923	0.08	0.27
Incentive per capita	41923	0.17	0.87
Budget revenue per capita	41923	0.14	0.18
Credit	41923	0.33	0.47
Credit per capita	41923	0.14	0.16
Assets 1	41923	0.32	0.46
Assets 3	41923	0.12	0.32
Assets 4	41923	0.19	0.39

(b) Description of variables and sources

Variable	Description	Sources
Innovation	Granted patents, utility models, and trademarks (yes/no)	KOSGEB
Firm age	Number of years passed since the firm's creation	KOSGEB
Firm size	Number of employees	KOSGEB
R&D	Investment in Research and Development (yes/no)	KOSGEB
Education	Average enrolment year	KOSGEB
Technology	Use of plc, cnc, or robots in production process	KOSGEB
ICT	Number of computers in the firm	KOSGEB
Export	Exporting activity (yes/no)	KOSGEB
Outsourcing	Use of external laboratories and/or acquisition of external technology (yes/no)	KOSGEB
Quality labels	Ownership of quality certificates and/or labels (yes/no)	KOSGEB
Entrepreneur's education	Average enrolment year of the manager	KOSGEB
Entrepreneurship index	Entrepreneurial capacity at NUTS 3 level	
Start-up	Firms that are less than 5 years old	KOSGEB
Electricity per capita	Per capita electricity consumption in manufacturing industry	SPO
Power capacity	Total capacity in power equipment	SPO
Location	Being located in an industrial park/zone and/or technological park (yes/no)	KOSGEB
OIZ size	Number of plots in organized industrial estates	SPO
Location quotient	Location quotient at NUTS 2 level $LQ = \frac{e_i/e}{E_i/E}$ where e_i = local employment in industry i e = total local employment E_i = reference area employment in industry i E = total reference area employment industry (%) ISIC Rev.2 4-digit level	TUBITAK
Incentives	Holding at least one incentive certificate (yes/no)	KOSGEB
Incentive per capita	Per capita amount of investments with incentive certificates	SPO
Budget revenue per capita	Consolidated budget revenue per capita	SPO
Credit	Use of private bank credits and/or loans (yes/no)	KOSGEB
Credit per capita	Amount of industrial, commercial, and tourism credits per capita	SPO
Assets	Level of net assets Cat. 1 corresponds to assets <50 billion TL Cat. 2 corresponds to assets 51–150 billion TL Cat. 3 corresponds to assets 151–300 billion TL Cat. 4 corresponds to assets >151 billion TL	KOSGEB

(c) List of indicators used for the Entrepreneurship Index

- Export per capita
- Export growth rate
- Export variety index
- Ratio of employers in total population
- Number of firms with quality labels and certificates (per 10,000 firms)
- Number of patents and utility models per 10,000 firms
- Number of new firms per 10,000 firms
- Firm turnover ratio (entry/exit)
- The share of joint stock corporations in total firms
- The share of agricultural sector in new firms
- The share of manufacture sector in new firms
- The share of service sector in new firms
- Amount of subventions per capita
- Cancel rate of incentive certificates
- Number of incentive certificates per 100,000 firms

Notes

1. Although several definitions of entrepreneurship exist in the economic literature, we refer here to an entrepreneur as 'someone who specializes in taking responsibility for and making judgmental decisions that affect the location, the form, and the use of goods, resources, or institutions' following the definition given by Hébert and Link (1989).
2. Given our measurement of innovative capabilities, the terms 'innovative capabilities', 'innovation activities', and 'innovation performance' are thus used interchangeably throughout the chapter.
3. We follow (Acs et al. 2004) and define entrepreneurial capital as the capacity for economic agents to generate new firms.
4. See (TÜSİAD 2002) for more details about the survey.
5. 'SME Strategy and Action Plan' and 'Industrial Policy for Turkey' in 2003, 'European Chart for Small Enterprises'.
6. The lack of longitudinal data does not seem to be a major problem, since the entry and exit of small Turkish firms is mostly conducted for the sake of tax corruptions. In fact, most of the small Turkish firms prefer to exit the market and to reenter with a new name and tax number in order to benefit from the tax incentives.
7. See Appendix table 7.1 for the sectoral distribution of the whole sample and the entrepreneurial firms. The definition and classification of the sectors according to their technology level are from Hatzichronoglou (1997).
8. We consider that a firm is technology-intensive if it uses either programmable logic controller (plc), numerical controller (cnc), or robots in its production process.
9. For the description and summary statistics of independent variables, see Appendix tables 7.3a and 7.3b.
10. The only two indicators where the start-up companies did not perform better were the capital size and the number of export countries, two areas where young firms risk struggling more than established ones. While access to capital is a major problem for all SMEs in Turkey, to find trade partners and new markets are processes that require some time and experience, advantaging hence the older firms versus the younger ones.
11. The indicators that contributed to the entrepreneurship index are listed in Appendix table 7.3c. For more details on the construction of the entrepreneurship index, as well as an exhaustive evaluation of regional inequalities in Turkey, see Pınarcıoğlu and Işık (2004).
12. In Turkey, in order to qualify for an investment incentive, it is necessary to obtain an incentive certificate before the investment is initiated. An investment must meet a minimum equity ratio of 20 per cent and minimum value of 600,000 YTL (US$420,000) for the developed regions, 400,000 YTL (US$285,000) for the normal regions, and 200,000 YTL (US$140,000) for priority development regions.
13. See Hoetker (2007) for a critical review of the use of probit models.
14. In what follows, marginal effects evaluated at the firms' mean values for continuous variables and the discrete change in the probability for dummy variables are reported rather than the raw coefficients.

15. Even though the specification with sector- and region-specific effects performs slightly better, we choose the last one given the risk of multicollinearity which could arise when we introduce the NIS variables at the regional level.
16. Only 857 firms are older than 40 years and 2502 firms (less than 5 per cent) have more than 66 employees.
17. The marginal effect of a change in both interacted variables is not equal to the marginal effect of changing just the interaction term. Moreover, the sign may be different for different observations. Hence, we cannot determine the statistical significance from the z-statistic reported, we must compute the cross derivative of the expected value of the dependent variable and the test for the statistical significance of the interaction effect must be based on the estimated cross-partial derivative. For more detailed explanation, see Norton, Wang, and Ai (2004).

References

Acs, Z. J., D. B. Audretsch, P. Braunerhjelm, and B. Carlsson (2004) 'The Missing Link: The Knowledge Filter and Entrepreneurship in Endogenous Growth', Discussion Paper 4783. London: CEPR.

Akçomak, I. S., and E. Taymaz (2004) 'Assessing the Effectiveness of Incubators: The Case of Turkey', Working Paper 0412. Ankara: Middle East Technical University.

Audretsch, D. B., and M. Keilbach (2004) 'Entrepreneurship Capital and Economic Performance', *Regional Studies*, 38 (8): 949–59.

Audretsch, D. B., and R. Thurik (2001) 'Linking Entrepreneurship to Growth'. Working Paper 2001/2. Paris: OECD Directorate for Science, Technology and Industry.

Baroncelli, E., C. Kink, and B. S. Javorcik (2004) 'The Global Distribution of Trademarks—Some Stylized Facts', Policy Research Working Paper 3270. Washington, DC: World Bank.

Bascavusoglu, E. (2008) 'Impact of SME Policies on Innovation: The Turkish Case', Open Discussion Papers in Economics 70. Milton Keynes: Open University.

Blanchflower, D. G. (2000) 'Self-employment in OECD Countries', *Labour Economics*, 7 (5): 471–505.

Carree, M., and A. Thurik (2003) 'The Impact of Entrepreneurship on Economic Growth'. In Z. Acs and D. Audretsch (eds), *Handbook of Entrepreneurship Research*. Boston/Dordrecht: Kluwer Academic Publishers.

Carree, M., A. van Stel, R. Thurik, and S. Wennekers (2000) 'Business Ownership and Economic Growth in 23 OECD Countries', Discussion Paper 00-001/3. Amsterdam/Rotterdam: Tinbergen Institute.

Dahlman, C., and L. Westphal (1982) 'On Technological Effort in Industrial Development: A Survey'. In F. Stewart and J. James (eds), *The Economics of New Technology in Developing Countries*. London: Pinter.

Dosi, G. (1993) 'Technological Paradigms and Technological Trajectories: A Suggested Interpretation of the Determinants and Directions of Technical Change', *Research Policy*, 22 (2), 102–3.

Edquist, C., and B. Johnson (1997) *Systems of Innovation: Technologies, Institutions and Organizations*. London: Pinter.

European Commission (EC) (2003). 'European Innovation Scoreboard: Indicators and Definitions', Technical Paper 01. Brussels: EC.

——— (2005). 'European Trend Chart Report', Technical Report. Brussels: EC.

——— (2006). 'Screening Report on Turkey', Technical Report. Brussels: EC.

Fagerberg, J. (2003) 'Schumpeter and the Revival of Evolutionary Economics: An Appraisal of the Literature', *Journal of Evolutionary Economics*, 13 (2): 125–59.

Feldman, M. (1999) 'The New Economies of Innovation, Spillovers and Agglomeration: A Review of Empirical Studies', *Economics of Innovation and New Technology*, 8: 5–25.

Feldman, M., and D. B. Audretsch (1999) 'Science-based Diversity, Specialization, Localized Competition and Innovation', *European Economic Review*, 43: 409–29.

Fonseca, R., P. Lopez-Garcia, and C. A. Pissarides (2001) 'Entrepreneurship, Start-up Costs and Employment', *European Economic Review*, 45 (4–6): 692–705.

Freeman, C. (1987) *Technology Policy and Economic Performance: Lessons from Japan*. London: Pinter.

Glaeser, E., H. Kallal, J. Scheinkman, and A. Shleifer (1992) 'Growth in the Cities', *Journal of Political Economy*, 100: 1126–52.

Golden, W., E. Higgins, and S. H. Lee (2003) 'National Innovation Systems and Entrepreneurship', Working Paper 08. Galway: Centre for Innovation and Structural Change.

Greenhalgh, C., and M. Rogers (2007) 'Trademarks and Performance in UK Firms: Evidence of Schumpeterian Competition Through Innovation', Working Paper 300. Oxford: University of Oxford.

Griliches, Z. (1979) 'Issues in Assessing the Contribution of Research and Development to Productivity Growth', *Bell Journal of Economics*, 10 (1): 92–116.

——— (1991) 'Patent Statistics as Economic Indicators: A Survey', *Journal of Economic Literature*, 28: 1661–707.

Grossman, G. M., and E. Helpman (1991) *Innovation and Growth in Global Economy*. Cambridge, MA: MIT Press.

Harding, R., and N. Bosma (2007) *GEM 2006 Global Summary Results*. Babson, MA: Babson College and London: London Business School.

Hatzichronoglou, T. (1997) 'Revision of the High-technology Sector and Product Classification', STI Working Paper 1997/2. Paris: OECD.

Hébert, R., and A. Link (1989) 'In Search of the Meaning of Entrepreneurship', *Small Business Economics*, 1 (1): 39–49.

Hoetker, G. (2007) 'The Use of Logit and Probit Models in Strategic Management Research: Critical Issues', *Strategic Management Journal*, 28 (4): 331–43.

Kirzner, I. M. (1973) *Competition and Entrepreneurship*. Chicago and London: The University of Chicago Press.

Lall, S., U. Deichmann, and S. Coulibaly (2007) 'Urbanization and Productivity: Evidence from Turkish Provinces over the Period 1980–2000', Policy Research Working Paper 4327. Washington, DC: World Bank.

Lucas, R. E. (1988) 'On the Mechanics of Economic Development', *Journal of Monetary Economics*, 22: 3–42.

Malerba, F., and L. Orsenigo (1997) 'Technological Regimes and Sectoral Patterns of Innovative Activities', *Industrial Corporate Change*, 6: 83–117.

Malmberg, C. (2005) 'Trademark Statistics as Innovation Indicators?—A Micro Study', Working Paper 2005-17. Lund: Lund University, CIRCLE.

Mendonca, S., T. S. Pereira, and M. M. Godinho (2004) 'Trademarks as an Indicator of Innovation and Industrial Change', *Research Policy*, 33 (9): 1385–404.

Napier, G., S. S. Serger, and E. W. Hansson (2004) 'Strengthening Innovation and Technology Policies for SME Development in Turkey', Policy Report. Malmö: IKED.

Nelson, R., and S. Winter (1982) *An Evolutionary Theory of Economic Change*. Cambridge, MA: Belknap Press of Harvard University Press.

North, D. (1990) *Institutions, Institutional Change and Economic Performance*. Cambridge: Cambridge University Press.

Norton, E. C., H. Wang, and C. Ai (2004) 'Computing Interaction Effects and Standard Errors in Logit and Probit Models', *Stata Journal*, 4 (2): 154–67.

OECD (2002) *Small and Medium Enterprise Outlook*. Paris: OECD.

——— (2003) *Economic Survey of Turkey*. Paris: OECD.

——— (2004) *Small and Medium-Sized Enterprises in Turkey: Issues and Policies*. Paris: OECD.

——— (2005) 'A Framework for the Development and Financing of Dynamic Small and Medium Sized Enterprises in Turkey', Project Report. Paris: OECD Istanbul Centre for Private Development and Istanbul Stock Exchange.

Ozaslan, M., B. Dincer, and H. Ozgur (2006) 'Regional Disparities and Territorial Indicators in Turkey: Socio-economic Development Index (SEDI)', Conference Papers ersa06p858. Louvain-la-Neuve: European Regional Science Association.

Pınarcıoğlu, M., and O. Işık (2004) *Yeni Kalkinmacilik: Bolge Kalkinmada Yeni Arayislar*. EU-UNDP/GIDEM.

Porter, M. E. (1990) *The Comparative Advantage of Nations*. New York: Free Press.

Radosevic, S. (2007) 'National Systems of Innovation and Entrepreneurship: In Search of a Missing Link', Working Paper 73. London: Centre for the Study of Economic and Social Change in Europe.

Reeves, T. (2005) 'Turkey's Regional Policy on the Road to the EU', *Turkish Policy Quarterly*, 41 (4): 1–13.

Romer, P. M. (1986) 'Increasing Returns and Long-run Growth', *Journal of Political Economy*, 94 (5): 1002–37.

Romijn, H., and M. Albaladejo (2002) 'Determinants of Innovation Capability in Small Electronics and Software Firms in Southeast England', *Research Policy*, 31: 1053–67.

Schmoch, U. (2003) 'Service Marks as Novel Innovation Indicators', *Research Evaluation*, 12: 149–56.

Schneider, F. (2005) 'Shadow Economies of 145 Countries all over the World: What do we Really Know?', Working Paper 2005-13. Basel: Center for Research in Economics, Management and the Arts.

Sgard, J., D. Unal, and Y. Zlotowski (2004) 'La Turquie au Tournant'. La Lettre du CEPII 2004-09. Paris: CEPII.

Taymaz, E. (2005) 'Are Small Firms Really Less Productive?', *Small Business Economics*, 25 (5): 429–45.

Taymaz, E., and S. Ozler (2004) 'Labor Market Policies and EU Accession: Problems and Prospects for Turkey', Working Paper 0405. Ankara: Middle East Technical University.

Turkish Industrialists' and Businessmen's Association (TÜSİAD) (2002) 'Entrepreneurship in Turkey', Technical Report. Ankara: TÜSİAD.

——— (2003) 'European Union and Turkey: Towards Economic Integration', Report 03. Ankara: TÜSİAD.

Turkish Statistical Institute (TSI) (2002) 'Census of Industry and Business Establishments', Technical Report. Ankara: Turkish Statistical Institute.

——— (2006) 'Established and Liquidated Companies and Firms', Technical Report. Ankara: Turkish Statistical Institute.

United Nations Industrial Development Organization (UNIDO) (2005) 'Capability Building for Catching-up', Industrial Development Report. Vienna: UNIDO.

World Bank (2004) 'Turkey Knowledge Economy Assessment Study', Technical Report. Washington, DC: World Bank

World Economic Forum (WEF) (2005) 'World Competitiveness Report 2004–2005', Technical Report. Geneva: WEF.

——— (2007) 'World Competitiveness Report 2006–2007', Technical Report. Geneva: WEF.

——— (2008) World Competitiveness Report 2007–2008, Technical Report. Geneva: WEF.

Part III

Institutions, Policies, and Incentives for Innovation

8

Innovation Policy, Entrepreneurship, and Development: A Finnish View

Otto Toivanen

8.1 Introduction

At least since the 1950s it has been recognized that innovation is central to economic growth. It has also been well understood that innovations both generate and rely on externalities. These twin facts, together with market failures most often thought to arise in the financing of innovation, serve as the basic justification for public sector actions which nowadays are jointly labelled innovation and research policies. These have their roots in the policies following the Second World War (e.g. Mowery 1995) and rose to prominence, for example, in the EU's Lisbon strategy.[1] R&D subsidies are both one of the most important tools of innovation policy in practice (Nevo 1998) and one that has theoretical justifications in the endogenous growth literature (e.g. Howitt 1999, Segerstrom 2000). It therefore seems worthwhile to consider the theory and practice of R&D subsidy policy in some detail. A key characteristic of R&D subsidies—one that is of great importance for both fostering entrepreneurship and implementing this policy in a developing country context—are that they are a policy tool which in practice requires active decision-making. That is, once a proposal has been received, the agency administering the programme has to make an active decision on whether or not to grant a subsidy, and if so, what the amount would be. This is unlike the other popular tool for enhancing R&D investments of firms, namely tax credits. It is my view that while the theoretical foundation for R&D subsidies is sound, the problems with implementation may call into question the justifiability of such policies.

8.2 Theoretical justifications for innovation policy

Externalities are the key justification for active government policy also in relation to innovative activities. As long as the benefits from innovative activities are not completely captured by whoever conducted them, there is a potential wedge between social and private benefits. As an example, think of a new good. As long as the firm developing the product does not capture the whole social surplus the new product generates, there is room for government policy. The empirical literature on new goods typically finds large estimated consumer surpluses from new innovations.[2]

Knowledge spillovers to other innovators are another externality commonly linked to R&D and used as a justification for active government policy. The seminal paper of Jaffe (1986) found that other firms' R&D affected the productivity of firm R&D positively. In later work Jaffe (1989) found that university R&D has a positive direct effect on commercial innovation as measured by patents at least in some industries, and a positive indirect effect by increasing industrial R&D. Later Jaffe, Trajtenberg, and Henderson (1993) looked at patent citations and found that citations from later patents to earlier patents are greatly affected by geography, whereby a new patent is more likely to cite a patent that was granted to an applicant who is geographically close to the new applicant.

Market failure in supply of finance to innovation has provided motivation for various policies aimed at supporting private sector R&D. Notice that this is a market failure that comes on top of any product market externalities. Financial market imperfections thus potentially provide an additional and independent motivation for government intervention. Himmelberg and Petersen (1994) put this hypothesis to the test by studying the sensitivity of R&D investments to cash flow. Using data from small US firms they find evidence in favour of financial constraints. Hyytinen and Toivanen (2005) find evidence that in Finland the availability of government support affects firms most in industries that are relatively dependent on external financing.

Finally, several of the above mentioned market imperfections may take added relevance in the case of so-called general purpose technologies (GPTs) (Bresnahan and Trajtenberg 1995). Often cited examples of GPTs are steam, electricity, and semiconductors. GPTs give an extra role to innovation policy in that they easily feature 'multiple equilibria'. This means that in practice, if GPTs are important, to push a country from a 'low activity equilibrium' to a 'high activity equilibrium' may be possible with good policy.

The above discussion may lead one to think that there is plenty of scope for government action. Any identified market failure calls for government intervention of some sort. This optimistic view may be unwarranted: such a conclusion makes strong demands on a government's ability to design policies that would rectify identified market failures. One has to take into account the government's ability to deliver, too.

In real life, governments face informational constraints that may be as severe or more severe than those of firms. Modern economics has analysed markets that are *interim efficient*.[3] In such a case, a benevolent social planner could not improve on the market outcome, if subject to the same informational problems as the market. From this vantage point, many phenomena are no longer market failures that call for government action.

The government has one unique advantage over the private sector, though: it is in a position in that it need not necessarily cover the costs of information gathering and dissemination by the revenues so generated. This means that an active government may in some circumstances be able to achieve better than interim efficient outcomes.

The foregoing example raises an important point, namely the heterogeneity of the innovation policy environment. Firms, R&D projects, and innovations are highly heterogeneous. This means that a policy that is optimal in the strict sense of achieving Pareto efficiency should vary not only from firm to firm, but from project to project. This is what puts a government agency running a R&D subsidy programme under considerable informational requirements.

8.3 Innovation in Finland

R&D and innovation are important for the Finnish economy; as an example of the plentiful evidence, Jalava and Pohjola (2008) estimate that Finnish growth without 'the success of information and communication technologies (ICT) manufacturing' would have been 0.9 percentage points lower during 1990–2004.

R&D activities in Finland are characterized by active government involvement and planning on the one hand, and dominance of a few large firms on the other hand. Three key facts provide a way to summarize Finland from the point of view of innovation policy, and these are provided by Figures 8.1 and 8.2, and Table 8.1. Figure 8.1 shows the aggregate inputs into R&D as a percentage of GDP. Finland exhibits a consistent upward trend, Sweden being the country that invests most in R&D in relation to its GDP. It is also clear from the Figure that the R&D to GDP ratios of developing countries are significantly lower than those of developed countries.

Figure 8.2 shows the number of scientific publications per 1,000 inhabitants. Finland produces a relatively large number of publications for its size, some 30 to 40 per cent more publications per inhabitant than, for example, Norway, a country of similar characteristics. The first message from Table 8.1, exhibiting statistics on US patents, is the one pointed out in Trajtenberg

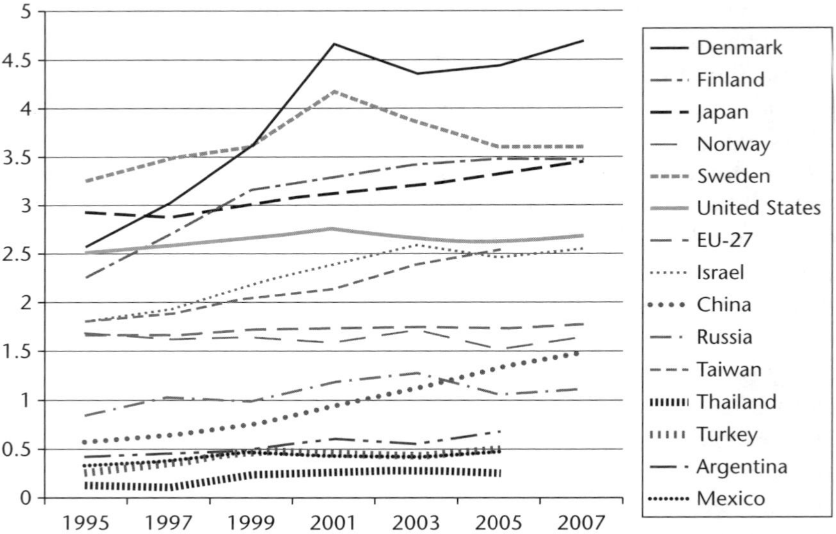

Figure 8.1 R&D as a share of GDP, selected countries

Source: Europe, EU27, Japan, and USA from <www.research.fi>, other countries from Szirmai (<www.dynamicsofdevelopment.com>).

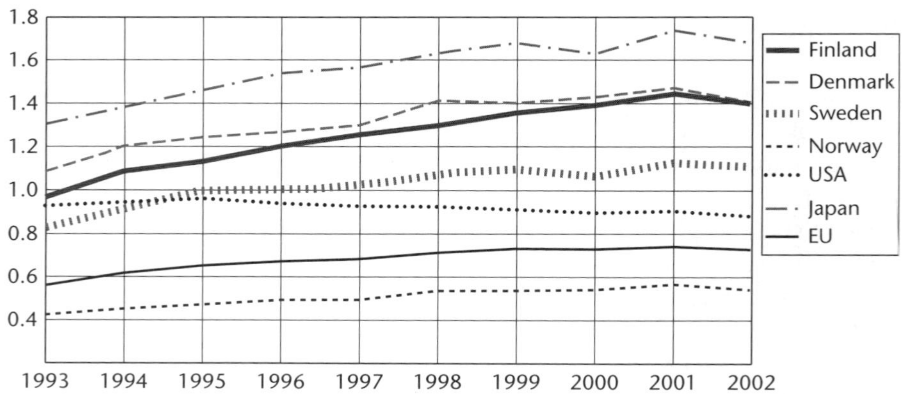

Figure 8.2 Scientific publications per thousand inhabitants, selected countries

Source: <www.research.fi>.

(2001): what matters in the end is not how R&D-intensive or successful one is in relative terms, but in absolute terms. This is especially true in the case of technologies where patents are an effective way to protect intellectual property (such as in the pharmaceutical industry). Taking this into account, Table 8.1 shows that Finland is doing very well considering its size. Israel and Finland—along with some of the Asian Tigers—belong to the small group of

Table 8.1 Figures on US patenting, various countries

Country	Patents per year		Patents per capita		Success rate in %		Annual growth rate in %	
G7	1968–97	1992–7	1968–97	1992–7	1968–97	1992–7	1968–97	1992–7
Canada	1,525	2,401	6.1	8.1	56	55	3.40	5.50
France	2,423	2,896	4.5	5.0	66	63	1.90	0.50
Germany	6,338	7,250	9.8	8.9	65	63	2.30	2.40
Italy	937	1,197	1.7	2.1	59	58	2.80	0.40
Japan	13,226	23,847	11.5	19.0	65	61	8.40	2.80
UK	2,547	2,494	4.4	4.3	55	51	0.20	3.10
USA	46,913	66,325	19.8	25.2	62	59	1.60	5.30
Reference Group								
Finland	214	438	4.5	8.6	57	58	8.60	10.00
Ireland	35	60	1.0	1.7	49	48	6.80	5.50
Israel	234	577	5.3	10.2	54	56	10.10	13.30
New Zealand	42	61	1.3	1.7	42	42	4.90	16.90
Norway	101	137	2.4	2.9	–	–	4.93	4.26
Spain	105	173	0.3	0.4	49	50	4.20	3.10
Asian Tigers								
Hong Kong	39	95	0.7	1.5	49	46	12.50	9.60
Singapore	22	83	0.8	2.6	55	52	16.50	10.30
South Korea	443	1,989	1.1	4.4	61	62	27.70	27.90
Taiwan	554	2,006	2.8	9.3	44	47	33.80	15.70

Source: Trajtenberg (2001) and author's own calculations using the NBER patent data for Norway.

countries that have been able to significantly improve their innovative output as measured by US patents.

The overall picture that emerges is pointed out by Trajtenberg (2001): Finland has succeeded in transforming itself by increasing its R&D inputs to a level seen only in a few other countries, in raising the output of the innovation sector to relatively good levels both in terms of scientific (academic publications) and commercial (patents) output.

8.4 R&D subsidies in Finland

8.4.1 *The Finnish innovation policy environment*

In this section, the overall structure of institutions of innovation policy in Finland as they now stand is briefly described. In a recent evaluation of the Finnish innovation support system, Georghiu et al. (2003) provide a description and also a short overview of the historical development of the system. The Finnish government—as many other governments—employs several agencies to conduct innovation policy. The most important ones are depicted in Figure 8.3.

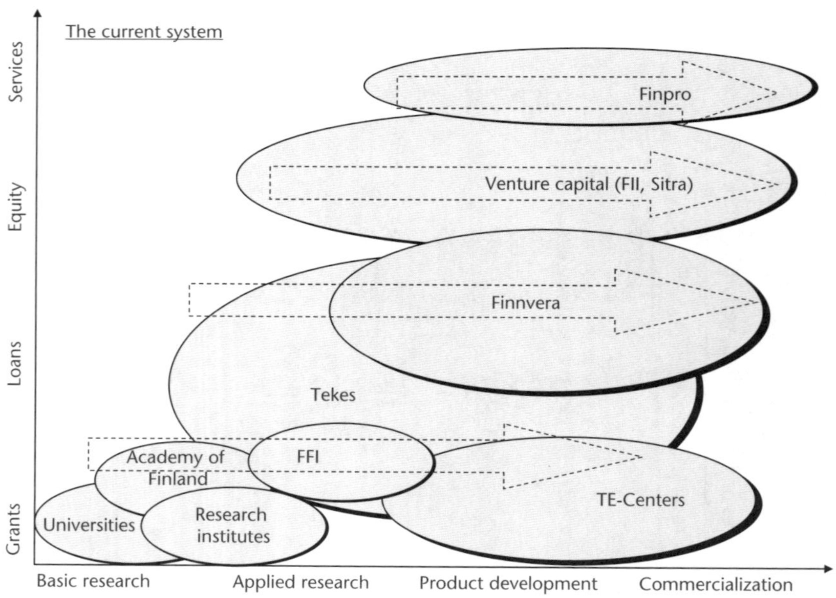

Figure 8.3 The Finnish innovation support system
Source: Georghiu et al. (2003).

The Academy of Finland is the main source of government funding for basic research, but it nowadays also funds applied research. The government runs a number of research institutes. Of these, the most important one is VTT, the Technical Research Institute. The Finnish Foundation for Innovation's objective is to help individuals and small and medium-sized enterprises (SMEs) in protecting their intellectual property. The regional dimension is taken care of through 15 TE centres. These are regional government offices whose task is to provide business support services, consultation, and advice, as well as finance to SMEs. The National Technology Agency of Finland (Tekes) is the main organization of Finnish innovation policy. Tekes provides funding (e.g. it is the sole source of R&D subsidies), expert advice, and promotion of national and international networking. Finnvera is a state-owned financing company whose main tasks are to promote entrepreneurship, development of SMEs, internationalization and exports of firms, and government regional policy. Only a comparatively small part of Finnvera's activities fall under innovation policy. Finnish Industry Investment and Sitra are the venture capital funds of the government, although the latter has lately greatly reduced its activities in this field. Finally, Finpro is the responsible organization for providing business support services for the internationalization of firms. Taken together, these organizations provide the Finnish government with a large set of tools to conduct innovation policy (for details, see Georghiu et al. 2003). Figure 8.4 shows the relative funding of the main innovation funding institutions—all of the above are not included because the share of innovation policy related to their remaining activities is small (e.g. Finnvera).

In Finland, the interaction between policy makers and policy-making bodies and interest groups is explicitly organized. This fact is reflected in the

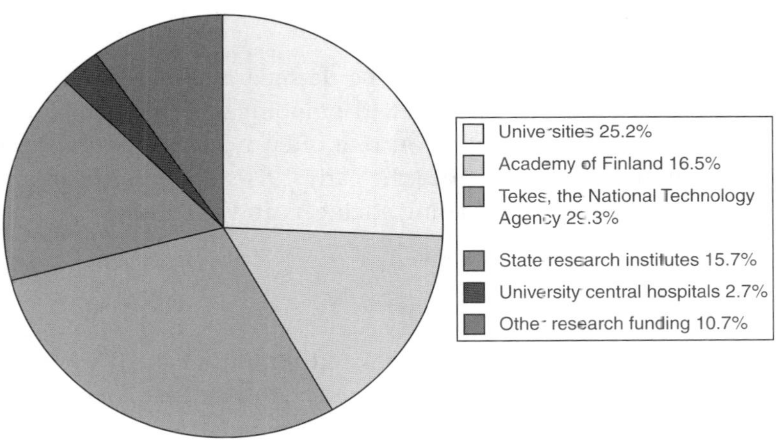

Figure 8.4 Relative funding of innovation policy institutions from Finnish state budget
Source: <www.research.fi>.

Table 8.2 Science and Technology Policy Council of Finland

Chairman
Prime Minister Matti Vanhanen
Deputy Chairmen
Minister of Education and Science Sari Sarkomaa
Minister of Economy Mauri Pekkarinen
Minister Members
Minister of Finance Jyrki Katainen
Minister of Agriculture and Forestry Sirkka-Liisa Anttila
Minister of Labour Tarja Cronberg
Minister of Health and Social Services Paula Risikko
Minister of Culture and Sport Stefan Wallin
Members appointed by the Government
Development Manager Eija Hietanen, The Central Organization of Finnish Trade Unions SAK
President and COO Olli-Pekka Kallasvuo, Nokia Corporation
Director General Erkki K. M. Leppävuori, Technical Research Centre of Finland VTT
Chief Executive Marja Makarow, European Science Foundation
President Markku Mattila, Academy of Finland
CEO Pekka Mattila, Finnzymes Ltd.
Director General Veli-Pekka Saarnivaara, Tekes—Finnish Funding Agency for Technology and Innovation
Rector Marja-Liisa Tenhunen, Central Ostrobothnia University of Applied Sciences
Professor Päivi Törmä, Helsinki University of Technology
Rector Keijo Virtanen, University of Turku

Source: <http://www.minedu.fi/OPM/Tiede/tiede-_ja_teknologianeuvosto/kokoonpano/neuvosto.html?lang=en> (accessed 4 October 2008).

Science and Technology Policy Council of Finland. The Council, chaired by the Prime Minister, is 'responsible for the strategic development and coordination of Finnish science and technology policy as well as of the national innovation system as a whole' (see <http://www.minedu.fi/tiede_ja_teknologianeuvosto/eng/index.html>). The current composition of the Council is given in Table 8.2. The Table shows the representation of the interest groups, as the Council members represent the most important ministries, the industry, the labour unions, and academia.

While Finland in many ways may be deemed a success in terms of its innovation policy, it is much harder to pinpoint what parts of the policy have been most crucial. To mention one often neglected feature (pointed out by Georghiu et al. 2003), since the early 1970s there has been a steady increase in the number of graduating engineers and scientists.

8.4.2 *Tekes*

Tekes is the main innovation policy tool and administers R&D subsidies. It is therefore worth describing in a little more detail how Tekes operates in this regard. Tekes' main objective is to promote the competitiveness of the Finnish industry and the service sector by providing funding and advice to both business and public R&D. Tekes decision-making is constrained by rules that

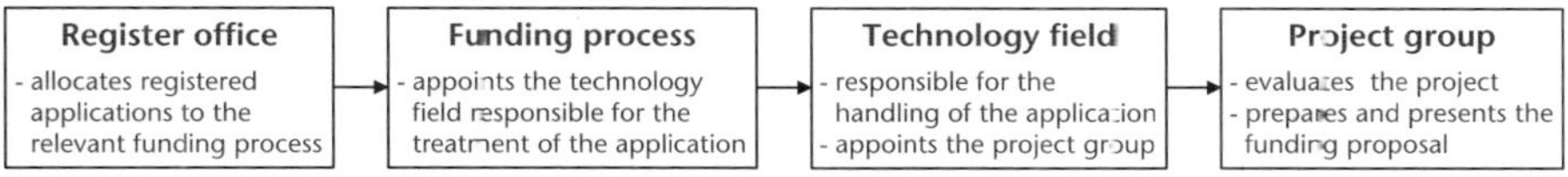

Figure 8.5 The Tekes decision process
Source: Tanayama (2007).

dictate that the maximum subsidy is 50 per cent of incurred costs (60 per cent for medium-sized and 70 per cent for small enterprises). In case of cooperation, these funding limits may be exceeded. Besides firms' R&D projects, Tekes funds feasibility studies and university research. Tekes receives some 3,000 applications per year, half of which are for business sector R&D projects, of which about two-thirds are accepted. Tekes grants around 300 million euros in subsidies and loans to companies. In 2007, Tekes funded 595 microenterprise projects with a total of 69 million euros; 408 small enterprise projects with a total of 73 million euros; 152 medium-sized enterprise projects with a total of 19 million euros; and 295 projects by large firms with 123 million euros (see <http://www.tekes.fi/en>). Tekes takes unsolicited applications, but also runs special programmes. The latter are usually designed in close cooperation with the industry.

Tekes has three funding instruments: grants (= subsidies), low interest loans, and capital loans. Low interest loans not only have a low interest rate, but are also soft: if the firm can demonstrate that the R&D project failed, it may waive the payment in part or completely. Capital loans are a Finnish speciality: they are included in fixed assets in the balance sheet, can be paid off only when the unrestricted shareholders' equity is positive and the debtor cannot give collateral.

Takalo, Tanayama, and Toivanen (2008) provide a description, a game theoretic model, and an empirical analysis of the application process. The application process goes as follows (see Figure 8.5): first, a firm makes an application. In the application, the firm describes the objectives and planned execution of the project, and a proposed budget. Firms are often in touch with Tekes officials prior to submitting the actual application. After receiving an application, Tekes screens it in a variety of dimensions. The decision-making body which decides both on the type and the amount of support varies depending on the amount of support (subsidy) to be decided upon as described in Table 8.3.

Table 8.3 Tekes decision-making limits

Amount related to the proposal	Decision-maker
≤ €150,000/200,000	Head of the technology field
≤ €500,000	Process leader
≤ €1,700,000	Director general
> €1,700,000	Board

Source: Tanayama (2007).

8.4.3 *Tekes and entrepreneurship*

Tekes is not the main Finnish policy tool to foster entrepreneurship (that is Finnvera), but it has currently specific policies that are aimed at newly established ventures. As reported above, quite a large number of microenterprise projects are funded by Tekes annually. Tekes (see <http://www.tekes.fi/en>) divides its special funding for young innovative companies into three stages: pre-stage funding, first stage funding, and second stage funding.

Pre-stage funding is aimed at helping at the planning stage. Support at this stage can be used to develop a business plan that is based on a high technology or knowledge-intensive concept.

First stage funding is available to those firms that have successfully completed the pre-stage, are small, less than 5.5 years old, and fulfil some other requirements. Those that are eligible have their proposals passed to a panel that consists mostly of venture capitalists. The role of the panel is to advise Tekes in decision-making.

In the second stage, only firms that are deemed to have a chance for rapid growth in international markets get funding. Funding is available for firms that are no more than eight years old, and funding is capped at €750,000.

While all Tekes programmes are evaluated, and Tekes commissions regularly research to assess the impact of its policies, I am not aware of any evaluations of the funding for young innovative companies as yet; this is probably explained by the fact that this programme is very recent.

8.5 R&D subsidies and entrepreneurship

As is hopefully clear from the above discussion, R&D subsidies are not directly aimed at promoting entrepreneurship in Finland—for this, there is an entirely different government agency. As is hopefully, however, also clear, Tekes does have policies directly aimed at start-ups and hence, fledgling entrepreneurs.

There are a number of obstacles that hinder the use of R&D subsidies in promotion of entrepreneurship. One is that subsidies in most countries are paid *ex post* against receipts. If it is the case that (new) entrepreneurs are credit constrained, this type of funding may be of little help. Second, while quite fast, the speed of the decision-making progresses in Tekes may be too slow for an entrepreneur. This may sound trivial, but is not at all; also, it may be hard to rectify. The reason for the former is that an entrepreneur may indeed need to move fast in order to e.g. secure funding for R&D. The reason for the latter is that the essence of R&D subsidies is that they are tailored, application by application. This tailoring is the greatest asset of R&D subsidies in comparison to R&D tax credits and almost by definition leads to an application process that may be too slow, especially for small firms.

Both of these aspects generate opportunity costs to entrepreneurs and, indeed, Takalo, Tanayama, and Toivanen (2008) find that the application costs of those who do not apply may be quite high. They also find that application costs are decreasing as the size of the firm increases (measured by the log of the number of employees)—a finding that suggests that entrepreneurs may be at a disadvantage from this point of view. An example of an application cost (for R&D subsidies) that is generated through opportunity costs is the increased risk for a (small) company not to be able to launch its product to the market place before its rivals—the potential loss of a first mover advantage. One should not underestimate the severity of such costs.

While thus a potential obstacle, the fact that R&D subsidies are tailored is also their biggest asset in terms of promoting entrepreneurship. If one is able (e.g. the legislation does not rule out the possibility) to explicitly target the support to certain types of firms, it may be possible to increase their likelihood of applying.

Another potential benefit of R&D subsidies is that the government may use them to generate externalities. The Finnish government (Tekes) does this by requiring that a large firm has to cooperate with SMEs in order to be eligible for government R&D support. This may allow small firms and entrepreneurs to access information that would not be available to them otherwise.

Finally, subsidies may act as a signal to private financiers. This can be the case if the government agency is good at finding out the type of the applicant—type here meaning 'good' (positive expected profits) or 'bad' (negative). Takalo and Tanayama (2010) provide a theoretical analysis of the circumstances under which the government may have such a 'certification' role.

Let us take as granted for the moment the need to foster innovation-oriented entrepreneurship. It could well be that R&D subsidies are not as good a policy tool for this purpose as some alternatives. Evidence seems to suggest that the hurdles to apply are high, the speed at which applications can be processed may be too slow, and opportunity costs of applying forbidding for those entrepreneurs who would be the target of such a policy. An alternative could be to offer R&D tax credits in such a way that even firms that do not make a profit can benefit from them. Examples are given by the Netherlands, where the tax credit is based on the (R&D) wage bill, and by Norway, where the tax credit is paid even if the firm is not profitable.

Such arrangements take R&D tax credits very close to subsidies. This is indeed the case, but there is one big difference: in the Netherlands and in Norway, the process through which one becomes eligible (e.g. in Norway there is *ex ante* screening for tax credits, see Møen and Haegeland 2007) is much less arduous than the one a firm needs to traverse in order to obtain R&D subsidies.

8.6 R&D subsidies and entrepreneurship in a developing country

It is clear that the Finnish policy described and discussed above is tailored to Finnish needs, and therefore reflects the fact that Finland is a developed economy with a highly educated workforce. One may therefore wonder how much of the Finnish experience can be of use in a developing country context. As there is to my knowledge no research on such issues, what follows should be viewed as speculative.

A key characteristic of the Finnish innovation policy system is that it has been quite predictable. It is clear that for any system to take root, information about it and its workings needs to spread among potential applicants. This may still be a problem in Finland; Tekes has existed for a quarter of a century, is well-known, and receives a lot of media attention, but still only a relatively small proportion of Finnish companies apply for funding from Tekes. At any rate, however, it would seem that a key characteristic to copy in a developing country is the long-term nature of the institutional arrangement.

A prerequisite of being able to administer an R&D subsidy programme is that the required human capital exists. One may copy institutional arrangements, but the institutions need to be manned by competent personnel. A key is to be able to recruit capable reviewers. Unless one succeeds in this, there is no hope of being able to reap the benefits that the system might otherwise offer.

A further 'capability requirement', a necessary condition really, is that there is a demand for such policy instruments. That most likely necessitates long-term investments in the education sector, from bottom to top. It seems fair to state that a key feature of the Finnish innovation policy has been a determined effort to increase the educational level of the population, and especially, the investments in engineering education. This is reflected, for example, in the fact that some 80 per cent of Finns obtaining a US patent and trademark office patent are engineers by training (Toivanen and Väänänen 2008).

There is also a danger that one should not forget: while the public sector intervention above was motivated by the possibility of the public sector to be a focal point, this same argument may also carry a danger. If the government sector is successful in becoming the focal point of R&D support, this may paralyze private sector efforts. In practice, it could be that e.g. private venture capital markets are not born if the government crowds them out. There is to my knowledge no evidence of this having happened anywhere in general or in

Finland in particular, but certainly this is an issue discussed by Finnish researchers regarding the Finnish R&D funding mechanisms.

While the message from the above analysis may be bleak, there is some evidence that public support to private R&D works (also) in developing countries. Hall and Maffioli (2008) analyse the effects of technology development funds in Argentina, Brazil, Chile, and Panama, and find encouragingly large positive effects on R&D investments, but less encouragingly little impact on patents, new product introduction, or productivity. One should also note that most of the countries studied by Hall and Maffioli differ from many developing countries in that they have a long history of investing (in relative terms) in higher education and research. Certainly it is the case that the Finnish experience is the more relevant, the better educated the workforce, and the better the current governance mechanisms of the government. Thus, one could imagine that Latin American countries had more to learn from the Finnish experience than most Sub-Saharan countries.

8.7 Conclusions

Innovation policy in general and R&D subsidies in particular are believed to be important in fostering economic growth. These beliefs are both held by politicians and backed by economic research. It is standard for researchers (e.g. Klette and Møen 1999) to voice concerns over political economy issues. These may lead to a situation where actual policies diverge from optimal ones. Government failure is the rubric under which one generally collects such problems. A potentially even more fundamental problem is that one should not use failure to be at a Pareto optimum as a justification for government intervention. A more reasonable, yet still challenging, yardstick would be to take seriously the informational constraints of the government when analysing the need for existing or new policies.

An important insight that can be derived even from the simplest model of R&D subsidies is that optimal subsidies are heterogeneous, i.e. they ought to vary from firm to firm and from project to project, and that the informational demands for designing optimal policies are potentially prohibitive.

The description of the Finnish innovation policy reveals that it is at the same time overarching and pays attention to small details e.g. when screening R&D subsidy applications. The policy towards entrepreneurship has to a great extent been separated from innovation policy, although at least recently one can see that innovation policy is being engineered also towards newly established small firms.

While the Finnish innovation support system seems to function relatively well and the R&D subsidy part especially so, it is still an open question what their role is in Finland's transformation into an (more) innovation-driven economy. In particular, it is not clear to what extent Finnish institutional arrangements can be copied into other environments, and whether it makes sense to copy them piecemeal or not.

Another lesson to be taken from small developed and R&D-intensive countries such as Finland and Israel is that innovation-linked based entrepreneurship can be fostered. A prerequisite for this is that the educational system harnesses individuals with the requisite skills; then and only then can active government policies towards entrepreneurship have the desired consequences. Finland and Israel offer a contrast in this respect, as, at least as an approximation, the former is dominated by large companies in terms of innovation, the latter by small businesses. Small countries can overcome the deficit of size by determined efforts especially in higher education.

Entrepreneurship, and especially entrepreneurship in connection to innovation, seems to hold the promise of economic prosperity and growth. Some characteristics of the Finnish policies, such as tying support for large companies to a requirement that they cooperate with small companies, may indeed be worth copying into a developing country environment. Others, such as relying entirely or mostly on R&D subsidies, may be less promising. As so often, the devil is in the details.

Notes

1. See, e.g., <http://europa.eu.int/comm/enterprise/innovation/communication.htm>.
2. Examples of studies with such findings are Trajtenberg's (1989) analysis of computerized tomography scanners, and Petrin's (2002) analysis of the introduction of minivans in the USA.
3. This concept was introduced by Holmström and Myerson (1983).

References

Bresnahan, T., and M. Trajtenberg (1995) 'General Purpose Technologies: 'Engines of Growth'?', *Journal of Econometrics*, 65: 83–108.

Georghiu, L., K. Smith, O. Toivanen, and P. Ylä-Anttila (2003) 'Evaluation of the Finnish Innovation Support System', Finnish Ministry of Trade and Industry Publications 5/2003. Helsinki: Finnish Ministry of Trade and Industry.

Hall, B., and A. Maffioli (2008) 'Evaluating the Impact of Technology Development Funds in Emerging Economics: Evidence from Latin-America', *European Journal of Development Research*, 20: 172–98.

Himmelberg, C. B., and B. Petersen (1994) 'R&D and Internal Finance: A Panel Study of Small Firms in High-tech Industries', *Review of Economics and Statistics*, 76 (1): 38–51.

Holmström, B., and R. Myerson (1983) 'Efficient and Durable Decision Rules with Incomplete Information', *Econometrica*, 51: 1799–820.

Howitt, P. (1999) 'Steady Endogenous Growth with Population and R&D Inputs Growing', *Journal of Political Economy*, 107: 715–30.

Hyytinen, A., and O. Toivanen (2005) 'Do Financial Constraints Hold Back Innovation And Growth? Evidence on the Role of Public Policy', *Research Policy*, 34: 1385–403.

Jaffe, A. (1986) 'Technological Opportunity and Spillovers of R&D: Evidence from Firms' Patents, Profits, and Market Values', *American Economic Review Papers and Proceedings*, 76: 984–1001.

——— (1989) 'Real Effects of Academic Research', *American Economic Review*, 79: 957–70.

Jaffe, A., M. Trajtenberg, and R. Henderson (1993) 'Geographic Localization of Knowledge Spillovers as Evidenced by Patent Citations', *Quarterly Journal of Economics*, 108: 577–98.

Jalava, J., and M. Pohjola (2008) 'The Roles of Electricity and ICT in Economic Growth: Case Finland', *Explorations in Economic History*, 45 (3): 270–87.

Klette, T., and J. Møen (1999) 'From Growth Theory to Technology Policy—Coordination Problems in Theory and Practice', *Nordic Journal of Political Economy*, 25: 53–74.

Møen, J., and T. Haegeland (2007) 'Input Additionality in the Norwegian R&D Tax Credit Scheme', Report 47. Oslo: Statistics Norway.

Mowery, D. (1995) 'Survey of Technology Policy'. In P. Stoneman (ed.), *Handbook of the Economics of Innovation and Technical Change*. Oxford: Basil Blackwell.

Nevo, R. (1998) 'Trends and Patterns of Public Support to Industry in the OECD Area', *STI Review*, 21: 12–14.

Petrin, A. (2002) 'Quantifying the Benefits of New Products: The Case of the Minivan', *Journal of Political Economy*, 110: 705–29.

Segerstrom, P. (2000) 'The Long-run Growth Effects of R&D Subsidies', *Journal of Economic Growth*, 5 (3): 277–305.

Takalo, T., and T. Tanayama (2010) 'Adverse Selection and Financing of Innovations: Is there Need for R&D Subsidies?', *Journal of Technology Transfer*, 35 (1): 16–41.

Takalo, T., T. Tanayama, and O. Toivanen (2007) 'Selection or Self-rejection? Applications into a Treatment Program: The Case of R&D Subsidies', Discussion Paper 76. Helsinki: HECER.

——— ——— ——— (2008) 'Expected Benefits of Innovation Policy', mimeo. Helsinki: HECER.

Tanayama, T. (2007) 'Allocation and Effects of R&D Subsidies: Selection, Screening and Strategic Behavior', Doctoral Dissertation. Helsinki: Helsinki School of Economics.

Toivanen, O., and L. Väänänen (2008) 'Returns to Inventors', Discussion Paper 237. Helsinki: HECER.

Trajtenberg, M. (1989) 'The Welfare Analysis of Product Innovations, with an Application to Computed Tomography Scanners', *Journal of Political Economy*, 97: 444–79.

——— (2001) 'Innovation in Israel 1968–97: A Comparative Analysis Using Patent Data', *Research Policy*, 30: 363–89.

9

Promoting Knowledge-intensive Entrepreneurship in India

Sunil Mani

9.1 Introduction

The recent growth performance of India's economy has attracted a fair amount of attention from various constituencies. The country, which has been variously described as a great underachiever of sorts is now being regarded as a knowledge powerhouse well on the way to becoming an important player in the international technological arena. There is now considerable interest among researchers and policymakers to understand the real factors behind this spectacular economic achievement of the country. Although there is now a fair amount of consensus on the fact that this growth performance can be largely traced to the process of economic liberalization set into motion since 1991,[1] it is also equally agreed that India's private corporate sector has responded to the signals provided by the state in a very admirable way.

For instance, both the savings and investments of the private corporate sector have really shown significant increases in the period since 2003–4. The sector has become very dynamic and is in the forefront of enabling the globalization of India's economy. Indicators of this globalization include the following: (i) there has been a significant improvement in the average export intensity of an Indian private sector firm; it increased from about 8 per cent in 1991 to about 25 per cent in 2007; (ii) Indian firms have made a number of acquisitions abroad and as a result the ratio of outward to inward FDI now stands at around 0.61; and (iii) a number of knowledge-intensive firms have emerged and these firms have become important forces to be reckoned with in their respective global field of operations. These firms range from auto components to biotechnology to IT software, and to wind turbines (See Table 9.1).

Behind the success of each of these 'blue chip' companies is the hard work put in by an entrepreneur or a group of entrepreneurs. These 'entrepreneurial' firms are different from the conventional enterprises on a number of parameters. Based on three traits in particular, the 'entrepreneurial firms' stand out from 'conventional' firms. These are (i) corporate governance: the entrepreneurial firms although established by a specific, very often, technically trained entrepreneur, are listed public limited companies with a wide shareholding. Having been listed on both Indian and foreign stock exchanges they are subject to more transparent disclosure practices regarding their operations and performance; (ii) technology-intensive industries: almost all the entrepreneurial firms operate in technology-intensive industries and mostly in service industries where the entry barriers are low; and (iii): extent of globalization: most of the entrepreneurial firms are highly integrated into the global economy. Exports of these enterprises typically range between 30 to 95 per cent of its total sales.

The Indian private corporate sector, which did not have a good record during the license-permit Raj phase, is now emerging as a strong innovation-based powerhouse. While there are many factors contributing to this, the key to this success can be traced to successful technology-based entrepreneurship. This entrepreneurship has been nurtured by the emergence of a number of institutional mechanisms, the most important of which is venture capital. Although the absolute level of venture capital investments in India is low, it has been growing at a rate of 90 per cent over the last few years and at this rate of growth, the industry is supposed to match Europe by 2010. Nevertheless at present, most Indian companies still finance their growth and expansion through internal resources. A second contributing factor is the availability of technically trained personnel including those trained abroad and willing to return to their homeland to start technical ventures. Apart from the few famous cases of firms, whole industries such as information technology, biotechnology, and aerospace industries have been jump-started by the emergence of this knowledge-intensive entrepreneurship.

The emergence of innovation-based private entrepreneurship in India contrasts with the Chinese experience where state firms continue to play a significant or dominant role (Kroeber and Yao 2008).

The purpose of this chapter is to understand the growth of knowledge-intensive entrepreneurship in India. Further it identifies the main facilitating factors or the constraints to this process so that public policy can be applied to correct for this as the case may be.

9.2 Growing interest in the study of knowledge-intensive entrepreneurship

Entrepreneurship in general is receiving greater attention from policymakers and experts in developed and developing countries. New dynamic enterprises contribute to economic development in several ways: as an important channel to convert innovative ideas into economic opportunities, as the basis for competitiveness through the revitalization of social and productive networks, as a source of new employment, and as a way to increase productivity. The link between entrepreneurs and economic growth, theoretically speaking, looks reasonably straightforward: entrepreneurs create new businesses, and new businesses in turn create jobs, intensify competition, and may increase productivity through technological change. High levels of entrepreneurship will thus translate directly into high levels of economic growth. However, the reality is more complicated. It is important to distinguish between 'necessity entrepreneurship' and 'opportunity entrepreneurship'. In necessity entrepreneurship, one has to become an entrepreneur because there is no better option for the person involved, whereas opportunity entrepreneurship is an active choice to start a new enterprise based on the perception that an unexploited or underexploited business opportunity exists. Necessity entrepreneurship has little or no effect on economic growth while opportunity entrepreneurship has a positive and significant effect. Opportunity entrepreneurship will necessarily involve innovation.

There has been recent renewed interest in the study of entrepreneurship in India and in China. A number of new books have been published documenting the emergence and history of recent entrepreneurship in the country (Bansal 2008; Damodaran 2008; Karki 2008; Khanna 2008). While these are studies of Indian entrepreneurship in general, the whole issue of knowledge-intensive entrepreneurship is clearly unexplored. In fact the number of studies on this aspect elsewhere is also limited, but with the growth of new technology-based industries there is a renewed interest in the issue. For instance, the most important journal in the economics and policy studies of technological change, *Research Policy*, had a special issue devoted to this aspect.[2] The nine articles in the issue examined 'the effect of environmental conditions on technology entrepreneurship, the processes by which entrepreneurs assemble organizational resources and technical systems, and the strategies used by entrepreneurial firms to pursue opportunities. The papers drew upon a wide variety of empirical evidence, from large sample analyses of archival data to detailed qualitative investigations'. But all the evidence and discussions were with reference to the developed economies.

In the Indian context, Taube (2009) and Athreye (Chapter 11 in this volume), are among the few studies that have examined the emergence of entrepreneurship in the context of the rapidly growing IT industry. The focus of the Taube study is more on the analysis of geographical concentrations of the IT industry and the coevolution of supportive institutions. The main hypotheses that are explored are that education, venture capital, and socio-cultural factors such as ethnic and gender diversity influence the pattern of knowledge-intensive industries like software.

The emergence of the new technology-based industries such as IT and biotechnology (BT) has opened up a world of new opportunities for new companies which hitherto did not exist. Further it appears that the cost of entry to these new economy industries is considerably lower than in the old economy industries especially from the point of view of new and young entrepreneurs. It will thus be instructive to see if the knowledge-intensive entrepreneurship is on the increase in India since the liberalization process. In order to place the discussions of it in the broader context of improved growth performance of India's economy since 1991 and especially since 2000, we start by mapping out the background to our study.

9.3 The background

9.3.1 *Overall growth performance and contribution of various industrial sectors to this growth process*

There is now considerable national and international interest in the growth performance of India's macroeconomy.[3] Although the economy's growth (along with that of China's) is one of the highest in the world, much of this growth has actually emanated from the services sector (Table 9.1). However, both the industrial and the manufacturing sector within it have been growing almost at the same rate as the macroeconomy, especially since 2000–1. This implies that the liberalization measures with regard to the industrial sector have had a lagged positive effect on the growth performance of this sector. This lagged effect may be due to the fact that most of the liberalization measures were piecemeal, ad hoc, unstructured, and implemented in a haphazard way especially at the level of individual states and as a result it took quite a bit of time for them to percolate down.

Although the growth rate of the manufacturing sector has clearly started looking up since 2000–1, an analysis (Table 9.2) of the contribution of various individual industry groups shows that much of the growth during this period of high-growth was actually contributed by less technology-oriented industries such as basic goods and consumer durables, although the contribution of technology-oriented industries such as capital and intermediate goods have

Table 9.1 Overall growth performance of India's macroeconomy, 1990–2009

	1990–1 to 1999–2000 (average)	2000–1 to 2008–9 (average)	2005–6	2006–7	2007–8	2008–9
Agriculture and allied activities	3.2	2.8	5.9	3.8	4.5	1.6
Agriculture	3.3	na	6.1	3.8	na	na
Industry	5.7	6.5	8.0	10.6	8.1	2.6
Mining and quarrying	4.8	5.0	4.9	5.7	4.7	3.6
Manufacturing	5.6	7.1	9.0	12.0	8.8	2.4
Electricity, gas, and water supply	7.3	4.5	4.7	6.0	6.3	3.4
Services	7.1	9.0	11.0	11.2	10.7	9.4
Trade, hotels, transport, storage, and communication	7.5	10.2	11.5	11.8	12.0	9.0
Financing, insurance, real estate, and business services	8.1	8.7	11.4	13.9	11.8	7.8
Community, social, and personal services	6.5	6.4	7.2	6.9	7.3	13.1
Construction	5.6	10.2	16.5	12	9.8	7.2
Real GDP at factor cost	5.7	7.2	9.4	9.6	9.0	6.7

Source: Reserve Bank of India (2008a: 50).

Table 9.2 Weighted contribution* of various manufacturing industries to overall manufacturing sector's growth (percentage share)#

	Basic goods	Capital goods	Intermediate goods	Consumer goods	Total
1999–2000	27.5	10.1	37.4	24.2	100
2000–1	24.5	3.5	27.2	45.4	100
2001–2	31.5	−11.8	16.3	64.5	100
2002–3	27.4	16.2	19.3	37.0	100
2003–4	25.0	18.0	25.4	31.0	100
2004–5	21.3	16.0	20.2	42.9	100
2005–6	25.4	20.0	8.4	46.3	100
2006–7	27.2	17.6	27.0	28.5	100
2007–8	24.7	25.0	27.4	22.9	100

Notes: * Relative contributions are computed as the ratios (in percentage terms) of the change in the index of the respective industry group to the change in the overall index adjusted for the weight of the relative industry group.
#. The individual shares may not add up to 100 due to rounding.

Source: Reserve Bank of India (various issues).

actually shown significant increases over the last two years. It must, however, be stressed that the classification of industries such as basic and consumer goods as less technology-oriented and capital and intermediate goods industries as technology-oriented ones is somewhat arbitrary and not based on strict objective criteria.

9.3.2 *Knowledge-intensity of India's overall domestic production has increased*

One of the distinguishing aspects of India's growth performance especially since 2000 is that its knowledge-intensity has increased (for the empirical definitions see the notes to Table 9.3). In 2005–6 about 14 per cent of the overall net domestic product of the country derives from knowledge-intensive production.

Mirroring the general trend, much of the knowledge-intensive production comes from the services sector. Further the growth performance of the knowledge-intensive production sector is larger than that of the overall economy.

9.3.3 *Rising innovations in the Indian industrial sector*

Mani (2007) had already shown that the share of the industrial sector in the performance of R&D has doubled during the post-liberalization period and accounted for about 30.4 per cent of the overall gross expenditure on R&D in

Table 9.3 Share of knowledge-intensive production in India's overall domestic production

Fiscal year	NDP at factor cost	Knowledge-intensive manufacturing industry[1]	Knowledge-intensive services industry[2]	Knowledge-intensive production	Share of knowledge-intensive production
1	2	3	4	5 = (3+4)	6 = (5/2)*100
1999–2000	1,605,103	87,049	50,054	137,103	9
2000–1	1,670,448	92,256	66,880	159,136	10
2001–2	1,764,137	95,257	79,041	174,298	10
2002–3	1,824,635	99,760	96,196	195,956	11
2003–4	1,981,389	110,650	120,575	231,225	12
2004–5	2,126,018	125,795	149,060	274,855	13
2005–6	2,326,581	137,703	185,772	323,475	14
2006–7	2,549,648	153,787	100,492[3]	254,279[3]	n. a.

Note: [1] Knowledge-intensive manufacturing = Chemical and chemical products (24) + Metal products and machinery (28+29+30) + Electrical machinery (31+32) + Transport equipment (34+35); figures in parentheses indicate the NIC-98 codes of these industries;
[2] Knowledge-intensive services = Communication + Computer relating services + R&D services;
[3] Excludes communication services as CSO (2008) do not report this for 2006–7.

Source: Central Statistics Organization (2008).

Table 9.4 Patent applications by Indian public and private entities

Public enterprises/ organizations	IPO	USPTO	PCT	EPO	Total	Private enterprises	IPO	USPTO	PCT	EPO	Total
Council of Scientific Industrial Research (CSIR)	1523	356	381	240	2500	Ranbaxy Laboratories	320	108	458	194	1080
Indian Institutes of Technology	237	19	25	6	287	Dr Reddy's Laboratories	315	27	113	39	494
						Orchid Chemicals & Pharmaceuticals	149	17	47	11	224
Bharat Heavy Electricals	189	3	6	0	198						
Steel Authority of India	136	0	0	0	136	Cadila Healthcare	148	17	67	23	255
Defence Research & Development Organisation (DRDO)	83	3	11	4	101	Cipla	138	27	67	39	271
Indian Council of Agricultural Research	82	0	1	1	84	Larsen & Toubro Sun	123	2	2	0	127
Indian Space Research Organisation	67	1	1	1	70	Pharmaceutical Industries	121	18	81	12	232
Indian Institute of Science	51	3	13	5	72	TVS Motors	121	0	0	0	121
Total	**2368**	**385**	**438**	**257**	**3448**	Tata Steel	119	1	10	3	133
						Aurobindo Pharma	84	3	52	2	141
						Tata Motors	66	0	0	0	66
						Torrent Pharmaceuticals	54	4	20	9	87
						Lakshmi Machine Works	52	0	0	1	53
						Matrix Laboratories	43	3	47	10	103
						Total	**1853**	**227**	**964**	**343**	**3387**
Ratio of private to public	0.98										

Notes: IPO = Indian Patent Office; USPTO = US Patent and Trademark Office; PCT = Patent Cooperation Treaty; EPO = European Patent Office.

Source: Evaluserve (2008).

2005–6. Further the share of the private corporate sector in the performance of this R&D had also increased from about 40 per cent in 1985–6 to about 65 per cent in 2002–3. A similar picture is visible when one analyses innovative performance using data on patent applications (both at domestic and foreign patent offices), although here government research institutes under the Council of Scientific and Industrial Research network play an important role as well (see Table 9.4). The data on these conventional innovation indicators of R&D expenditure and patents applied for clearly show that the Indian private corporate sector's innovative performance has increased rather significantly during the period since economic liberalization, although this rising innovative performance is concentrated in certain specific industries (such as pharmaceutical) and in certain specific firms which are entrepreneurial in nature (for instance Dr Reddy's laboratories, Ranbaxi, Torrent, Orchid, and so on). In consequence, there is a change in the perception of India s private corporate sector from being *'bazaar style capitalists'* to entrepreneurial firms which are interested in improving their long-term competitiveness by investing in the creation of new technologies.

9.4 Growth of knowledge-intensive entrepreneurship in India

The term 'knowledge-intensive entrepreneurship' lacks a rigorous definition. The term is often used interchangeably with other terms such as technology entrepreneurship. In our frame of reference it means entrepreneurship in the context of medium- and high-technology industries, both in manufacturing and service. The medium- and high-technology industries that we consider are the following: chemical and chemical products, metal products and machinery, electrical machinery, transport equipment, communication services, computer relating services, and R&D services.

India's corporate sector, barring some notable exceptions, was not at all known for any major technology related activities until 1991. To a certain extent this somnolent nature of the corporate sector was attributable to the stifling external environment. It was characterized by a web of governmental regulations governing conditions of entry to the industrial sector, expansion and diversification of existing industrial activity, acquisition of technology from abroad, etc. All this was to change, albeit slowly, with the announcement of the new industrial policy statement of July 1991. According to Mohan (2006), 'massive deregulation of the industrial sector, in fact, constituted the first major package of reforms in July 1991'. The obsolete system of capacity licensing of industries was discontinued, the existing legislative restrictions on the expansion of large companies were removed, phased manufacturing programmes were terminated; and the reservation of many basic industries for investment only by the public sector was removed. At the same time restrictions on the import of foreign technology were withdrawn, and a new regime welcoming FDI, hitherto discouraged with limits on foreign ownership, was introduced. With this massive reform introduced in one stroke in 1991, the stage was set for a policy framework that encouraged new entry, introduced new competition, both domestic and foreign, which thereby induced the attainment of much greater efficiency in industry over a period of time. One area of industrial reform that has been sluggish has been the removal of restrictions that exist on investment in most labour-intensive industries—known as small-scale industry reservations. In 1991 as many as 836 industries were reserved for investment by only small firms, defined by the level of investment. The number of these industries has now come down to 326. It is now more or less accepted in the literature that from 1991 onwards the corporate sector in India has grown rapidly. In the following we discuss four macroindicators of this growth performance drawing essentially from a variety of official sources. Subsequently this is supplemented with some micro-level indicators of knowledge-intensive venture creations.

9.4.1 *Macroindicators*

9.4.1.1 GROWTH OF NEW VENTURE CREATION AND THE RELATIVE SIZE OF INDIA'S PRIVATE CORPORATE SECTOR

In order to measure this, we use two indicators: new company formation and the size of India's corporate sector in relation to its GDP (Table 9.5).

According to Table 9.5, the number of new companies formed has increased quite enormously from about 250,000 in 1992 to about 730,000 in 2006. On average about 34,000 new companies were established every year since 1992 although the rate of growth of new company formation has actually decelerated during the post-liberalization period. However, we do not have further data on whether these companies are started by new entrepreneurs or by existing entrepreneurs. But there is indirect evidence to show that most of the companies that have entered new technology-based industries such as IT, BT, and even the auto parts industries are new companies set up by hitherto not so well-known entrepreneurs.[4] As a result of this phenomenal growth of new companies the size of the corporate sector in India measured by the share of its paid-up capital to the country's GDP has increased by 12 percentage points to almost a fifth of its GDP by 2006—the latest year for which such data are available. Another interesting aspect is that the gross domestic saving and investment rates of the private sector have increased. For instance, the gross domestic savings rate of the sector has increased from 3.4 per cent in 2001–2 to 7.8 per cent in 2007–8 and the gross domestic investment rate has

Table 9.5 Trends in new company formation and the relative size of India's private corporate sector, 1980–2006

Year	New company formation	New company formation index	Paid-up capital as a per cent of GDP
1980	4,932	100	
1981	6,195	126	
1982	9,645	196	
1983	10,452	212	
1984	11,331	230	
1985	15,038	305	
1986	15,030	305	
1987	16,258	330	
1988	17,603	357	
1989	21,974	446	
1990	21,774	441	
1991	22,317	452	3.89
1992	25,896	525	3.94
1993	26,483	537	4.50
1994	28,758	583	4.83
1995	47,671	967	5.86
1996	55,833	1,132	6.78
1997	41,804	848	8.04
1998	33,547	680	8.42
1999	27,484	557	9.18
2000	30,428	617	10.36
2001	26,645	540	12.09
2002	20,151	409	12.86
2003	22,887	464	13.60
2004	29,331	595	14.44
2005	38,118	773	13.89
2006	52,496	1,064	17.31
Average number of new companies formed 1980–91	14,379		
Average number of new companies formed 1992–2006	33,835		

Source: Ministry of Company Affairs (2007).

increased from 5.4 to 14.5 per cent during the same period (RBI 2008a: 70–1). All this points to an improvement in entrepreneurial activity in the country.

9.4.1.2 GROWTH OF KNOWLEDGE-INTENSIVE VENTURES

For measuring this aspect we employ a direct measure and a proxy. The direct variable is the number of new company registrations in India according to the level of activity (National Knowledge Commission 2008). According to the National Knowledge Commission there are four levels of entrepreneurship in terms of the level of technology involved with low technology activities such as agriculture and allied activities at the bottom of the pyramid (Level 1) and knowledge-intensive sectors at the top of the pyramid (Level 4):

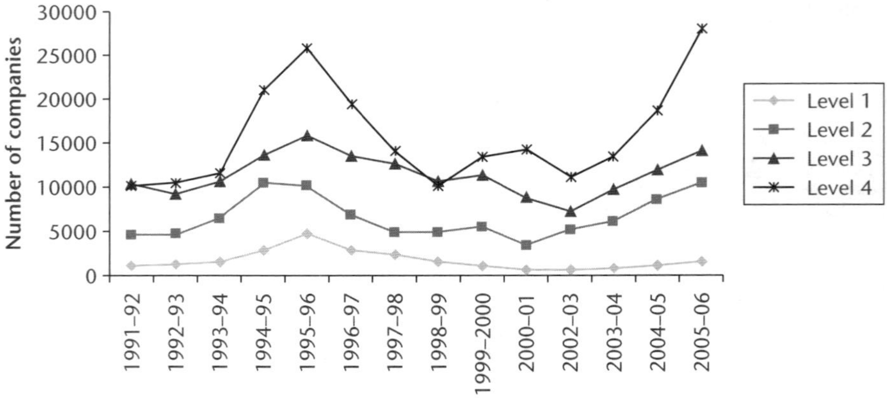

Figure 9.1 Distribution of new company formations in India according to intensity of knowledge

Source: National Knowledge Commission (2008: 6).

- Level 1. Agriculture and other activities: crop production, plantation, forestry, livestock, fishing, mining, and quarrying
- Level 2. Trading services: wholesale and retail trade, hotels and restaurants
- Level 3. Old economy or traditional sectors: manufacturing, electricity, gas and water supply
- Level 4. Emerging sectors (including knowledge-intensive sectors): IT, finance, insurance and business services, construction, community, social and personal services, supply chain, transport, storage, communications, etc.

The data on new company formations that we discussed in Table 9.3 could be cross-classified according to these four levels (Figure 9.1) and this shows that new companies belonging to knowledge-intensive sectors account for the largest share and the number of new companies formed has significantly increased since around 2003.

This dominance of technology-intensive sectors in total company formation is further corroborated by our proxy—namely the technology content of all industrial proposals actually implemented since 1991 (Table 9.6). Once again, with the exception of a few industries such as textiles, the majority of the new proposals are in technology-oriented industries such as chemicals, fuel, electrical equipments, etc. This, once again, suggests that technology-oriented ventures are on the rise in India since the initiation of economic reforms in 1991.

Table 9.6 New industrial ventures implemented

	Investments (in million Rs.)	Share (%)
Chemical other than fertilizer	378,690	14.09
Fuels	346,430	12.89
Metallurgical industries	303,960	11.31
Textiles	258,220	9.61
Prime movers*	232,910	8.67
Cement and gypsum	124,710	4.64
Electrical equipments	108,940	4.05
Others	104,390	3.88
Vegetable oil	73,960	2.75
Telecommunications	73,760	2.74
Leather, leather goods	70,780	2.63
Fermentation industries	65,870	2.45
Food processing industry	60,990	2.27
Sugar	59,000	2.20
Industrial machinery	57,960	2.16
Boilers and steam generators	38,920	1.45
Paper and pulp	38,380	1.43
Transportation industry	34,320	1.28
Rubber goods	32,200	1.20
Drugs and pharmaceuticals	29,950	1.11
Miscellaneous, mechanical industries	26,080	0.97
Machine tools	23,320	0.87
Glass	21,740	0.81
Glue and gelatine	19,230	0.72
Agricultural machinery	17,000	0.63
Scientific instruments	16,900	0.63
Ceramics	15,800	0.59
Soap, cosmetics, and toiletries	14,340	0.53
Commercial, HH equipments	10,430	0.39
Photographic raw film, paper	9,650	0.36
Fertilizers	6,450	0.24
Timber products	4,630	0.17
Miscellaneous industries	4,200	0.16
Medical and surgical instruments	1,950	0.07
Industrial instruments	950	0.04
Earth moving machinery	360	0.01
Dye stuffs	330	0.01
Total	2,687,730	100.00

Note: * This includes electricity generating equipment such steam and gas turbines, windmills, etc.

Source: Secretariat of Industrial Assistance (2008).

However, we do not have any data on the survival rates of these new ventures as it is quite possible that some of these may have exited from business due to a variety of reasons.

9.4.1.3 INDIAN INVESTMENTS ABROAD[5]

An increasing number of Indian companies are now investing abroad in order to access high-growth markets, technology, and knowledge, boost their positioning in the value chain, attain economies of size and scale of operations, tap global natural resource banks, and leverage international brand names for

Table 9.7 Ratio of outward to inward FDI in India, 1990–2008

	Inward FDI credit	Inward FDI debit	Net Inward FDI	Outward FDI Credit	Outward FDI debit	Net outward FDI	Ratio of outward FDI to inward FDI
1990–1	107	10	97				
1991–2	147	18	129				
1992–3	345	30	315				
1993–4	651	65	586				
1994–5	1,351	8	1,343				
1995–6	2,173	29	2,144				
1996–7	2,863	22	2,841				
1997–8	3,596	34	3,562				
1998–9	2,518	38	2,480				
1999–2000	2,170	3	2,167				
2000–1	4,031	0	4,031	70	829	759	0.19
2001–2	6,130	5	6,125	99	1,490	1,391	0.23
2002–3	5,095	59	5,036	73	1,892	1,819	0.36
2003–4	4,322	0	4,322	142	2,076	1,934	0.45
2004–5	6,052	65	5,987	35	2,309	2,274	0.38
2005–6	8,962	61	8,901	216	6,083	5,867	0.66
2006–7	22,078	87	21,991	881	14,393	13,512	0.61
2007–8	32,453	126	32,327	2,471	19,253	16,782	0.52

Source: Reserve Bank of India, Database on the Indian Economy, available at <http://dbie.rbi.org.in>.

their own brand building. Over time, net outward FDI from India works out to, on average, 42 per cent of net inward FDI to India (see Table 9.7).

An interesting point brought by Table 9.7 is that credit on account of outward FDI has been steadily increasing over the last three years and now works about US$2.5 billion—significantly above debits on account of outward FDI from India. This implies that investments made abroad by Indian companies are earning and dividends for them and the country profits, which when repatriated to India appear as a credit item on the balance of payments account. Most of these investments are in the manufacturing sector (Table 9.8), although in the most recent period the investments in trading have shot up.

Within manufacturing a number of technology-oriented industries such as pharmaceuticals, automobiles, basic metals, telecommunications, and

Table 9.8 Industry-wide distribution of FDI from India, 2004–8 (in million US$)

Industry	2004–5	2005–6	2006–7	2007–8
Manufacturing	1,170	3,407	3,545	6,240
Financial services	7	160	28	26
Non financial services	304	895	7,486	1,635
Trading	192	377	1,739	8,993
Others	100	207	656	1,010
Total	1,773	5,046	13,454	17,904

Note: Data include both equity and loan component.

Source: RBI (2008a: 154).

electrical equipments have been important. This increase in FDI from India has been facilitated by a number of favourable policy changes on the home front which encouraged such investments beginning with the Foreign Exchange Management Act of 1999.

In the above sections we have seen several related macroindicators of the growth of technology-based entrepreneurship in the country. We now present two case studies of technology-based entrepreneurship from the country. The two cases are widely discussed in the literature.

9.4.1.4 FINDINGS FROM THE GLOBAL ENTREPRENEURSHIP MONITOR[6]

Although GEM has been measuring the extent of entrepreneurial activity across the world since 1999, due to changes in data definitions and non-coverage of the Indian experience, we are constrained to present the data on entrepreneurial activity in India only for the most recent period of 2007. However, to interpret these figures we have tried to present two measures[7] of prevalence rates of entrepreneurial activity rates for the BRIC countries and United States in Table 9.9.

Although India has a lower value than both Brazil and China on the two prevalence rates, its score compares quite favourably with that of the USA especially with reference to the prevalence rate of overall entrepreneurial activity.

Further there is also a good amount of quantitative evidence that the so-called innovation ecosystem is becoming increasingly favourable. This was brought out by a recent study (KPMG-TiE 2008) measuring the 'Entrepreneurial Confidence Index' in ten states of India. Based purely on the perceptions of the entrepreneurs, rather than any factual analysis of the factors, the study aimed to identify the elements involved and benchmark the development of a conducive 'entrepreneurial ecosystem' across the country. The conclusions have highlighted the general confidence in the Indian economy and the belief that 'things are moving in the right direction'. In other words entrepreneurs

Table 9.9 Prevalence rates of entrepreneurial activity in BRIC countries and the USA, 2007

	Early stage entrepreneurial activity rate*	Overall entrepreneurial activity*	Number of observations
Brazil	12.7	22.4	2,000
China	16.4	24.6	2,666
India	8.5	13.9	1,601
Russia	2.7	4.3	1,939
United States	9.6	14.1	1,583

Note: * For definitions of these rates, please see note 7.

Source: Global Entrepreneurship Monitor (2007).

Table 9.10 Main features of the National Entrepreneurship Network start-ups

	Industry-wide distribution		Geographic spread			Year of establishment		
	Number	Percentage		Number	Percentage		Number	Percentage
IT/Internet/software	195	33	Bangalore	147	25			
Telecom/mobile	35	6	Mumbai	112	19	2003	35	6
Media/entertainment	56	10	Delhi (plus NCR)	110	19	2004	36	6
HR, recruiting, training, consulting, outsourcing	43	7	Chennai	52	9	2005	61	10
Education	42	7	Hyderabad	44	7	2006	118	20
Retail/consumer-based	33	6	Pune	42	7	2007	197	34
Hospitality/travel	24	4	Ahmedabad	15	3	2008	141	24
Finance	16	3	Kolkata	8	1	Total	588	100
Others	144	24	Jaipur	7	1			
Total	588	100	Coimbatore	7	1			
			Chandigarh	5	1			
			Indore	4	1			
			Other cities	35	6			
			Total	588	100			

Source: National Entrepreneurship Network Online <http://www.nenonline.org/>.

are bullish about the ecosystem. Expectations of entrepreneurs from the states like Gujarat, generally considered being leaders in entrepreneurship, expected more from their ecosystem and thereby held the state to a higher standard. The study reinforced the widely held assumptions that risk capital is still not available in the desired quantities; and governance issues and local environment in the ecosystem get low scores.

9.4.2 *Microindicators*

9.4.2.1 ANALYSIS OF RECENT INNOVATIVE START-UPS

An analysis of a unique dataset on entrepreneurship based on the nominees at the Tata-NEN hottest start-ups competition run by a not-for-profit organization, National Entrepreneurship Network (NEN),[8] has thrown up some additional insights into the emergence and growth of technology-based entrepreneurship in the country in recent times. Table 9.10 summarizes the main features of the sample entrepreneurs in terms of industry-wide, geographic spread, and year of establishment.

It is seen that approximately 40 per cent of the start-ups are technology-based (IT and telecoms), mostly based in the larger cities with a quarter of them in Bangalore itself and most of them established in the last three years. The earliest one in the sample was set up in 2003. This latter finding is quite consistent with our earlier finding in Table 9.3 that the real fillip to entrepreneurial activity took place only in the current millennium and specifically since 2004. The background of these new entrepreneurs also presents us with some interesting results (Table 9.11).

It is interesting to note that the majority of them are in their 20s, first generation entrepreneurs having their first business and having studied abroad or in Tier I institutions in India where they were exposed to the nitty gritty of starting a new business venture. Women formed only 8 per cent of the total number. In terms of head count, the total employee strength ranged from five to 15 employees.

9.4.2.2 DELOITTE TECHNOLOGY FAST 500 INDIA

Deloitte Touche Tohmatsu, one of the leading management consultancy firms, has been conducting a competition for the fastest growing technology company[9] in the Asia Pacific region known as the Deloitte Technology Fast 500 Asia Pacific.[10] In 2003 when the competition was started only 12 Indian companies made it into this ranking. However in 2007 there were 82 from India—and India was ranked number two in the top 500 with the largest set of high achievers. A run through this list showed that all the companies were new companies and more than half of them were in IT software. This again further substantiates the growth of this phenomenon.

Table 9.11 Background of the National Entrepreneurship Network start-ups

Age	Percentage	Family background	Percentage	Educational background	Percentage	Number of businesses	Percentage
Nominees in their early to mid-20s	25	Business	24	Studies abroad	17	First timer	65
Nominees in their late 20s	22	First generation	76	Tier I institutes	41	Serial entrepreneur	35
Nominees in their 30s	42	Total	100	Tier II institutes	40	Total	100
Nominees in their 40s	11			na	2		
Total	100			Total	100		

Source: National Entrepreneurship Network Online ⟨http://www.nenonline.org/⟩.

In sum, the macro- and microindicators that we have presented reinforce the point that the process of economic liberalization and international integration of India's economy has served to unleash a spate of entrepreneurship that was hitherto not seen or experienced in India's recent economic history.

9.5 Facilitating and constraining factors

The basic proposition that we have advanced so far, with the help of a variety of macro-and microindicators, is that there has been an increase in knowledge-intensive entrepreneurship in India since the onset of economic reforms in 1991. In this section, we will attempt to identify those facilitating factors and those which are still constraining a faster emergence of this activity. According to the GEM,[11] there are ten facilitating factors or framework conditions for this activity to flourish and sustain. These are: financial support, government policies, government programmes, education and training, R&D transfer, commercial, professional infrastructure, internal market openness, access to physical infrastructure, cultural, social norms, and intellectual property rights protection.

While all these factors are important, in the case of India, we can identify five key facilitating factors.[12] These are:

1. The new market opportunities presented by a liberalizing economy.
2. The availability of financial support schemes from both official and private sources.
3. The existence of a large number of governmental programmes and public–private partnership programmes.
4. The emergence of a number of private sector initiatives for supporting knowledge-intensive entrepreneurship by complementing government programmes and by reducing information asymmetries.
5. The increased availability of technically trained manpower due to a phenomenal increase in the enrolment rate for engineering and technology education at especially the tertiary level in the country.

9.5.1 *Growth in market opportunities*

An important aspect of liberalization that was set into motion in 1991 was that it pared down the discretionary role of the government with respect to economic matters and increased the scope of market forces. One of the important components of this increased space was the dispensing of the industrial licensing and other regulatory measures thereby reducing the height of barriers to entry for new entrepreneurs. This ease of entry, we argue, is one of the

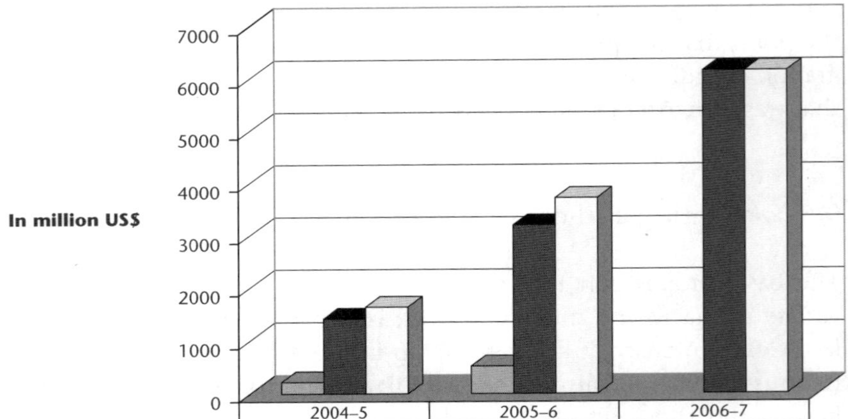

	2004–5	2005–6	2006–7
R&D services	221	519	
Architectural, engineering, and other services	1417	3193	6134
Total	1638	3712	6134

Figure 9.2 Growth of knowledge process outsourcing in India
Source: RBI (2006 and 2008b).

reasons for the rise of entrepreneurship in general (which was seen earlier in Tables 9.3 and 9.5). Against this background, an aspect that has promoted knowledge-intensive entrepreneurship is the emergence and growth of new technology-based industries such as IT and BT, which really opened up a new vista of opportunities. A run through the list of the top 20 enterprises (in terms of domestic and export sales) in each of these two industries show that almost all the enterprises were established during the 1990s or in the more recent period. Most of these are small and medium type enterprises initially set up by technology-oriented entrepreneurs. We further argue that a common factor in spurring opportunities in these two areas is the growth of knowledge process outsourcing. As can be seen from Figure 9.2, knowledge process outsourcings (proxied by receipts of R&D services, architectural, engineering, and other technical services) have been on the rise indicating further market opportunities in addition to the organic growth that is taking place in both the IT, BT, and other high-technology industries such as mobile telecommunications.

9.5.2 *Availability of finance and especially risk capital*

One does not have to emphasize the availability of financial resources, equity and debt, for new and growing firms including grants and subsidies. The availability of external risk capital has often been a constraining factor for financing company formation in India. Firms in India have increasingly relied on internal sources of capital and in terms of debt capital rather than equity for financing their long-term investment goals (Table 9.12). It is interesting to

Table 9.12 Sources of funds for private corporate sector manufacturing firms in India, 1995–2007

Item/Year	1995–6	1996–7	1997–8	1998–9	1999–2000	2000–1	2001–2	2002–3	2003–4	2004–5	2005–6	2006–7
Number of Companies	1930	1930	1948	1914	1927	2024	2031	2201	2201	2214	1064	1431
A. Internal sources of which	36.6	35.9	33.4	38.4	40.3	57.1	65.3	69.8	53.4	55.5	34.4	38.3
a) Reserves and surplus	20.8	16.4	11.2	8.8	9.1	6.7	–18.8	18	20	26.6	13.8	27.8
b) Depreciation	12.2	17.9	20.8	29.1	30.7	41.8	47.9	37.3	25.7	19.2	14.2	9.3
B. External sources	63.4	64.1	66.6	61.7	59.7	42.9	34.7	30.2	46.6	44.5	65.6	61.7
a) Paid up capital	13.9	10.1	7.6	11	21.9	12.8	10.5	9.4	9.3	10.8	22.8	12.8
Net issues	3.4	2.6	2.9	7.1	5.4	4.6	4.7	7.3	4.9	4.1	na	na
Premium	10.4	7.5	4.7	4	16.5	8.2	5.8	2.2	3.6	6.4	19.4	12.5
b) Borrowings of which	31.4	45.6	45.9	37.5	20.1	9.3	8.8	5.6	17	15.3	25.8	31.0
– Debentures	3.5	5.4	12.2	5.1	3.8	9.5	–1.5	–5.6	–3.5	–1.1	–4.1	–1.1
– From banks	17.7	13.3	10.1	29.3	8.4	–0.8	21.5	27.1	21.4	15.2	30.7	22.8
– From FIs	6.1	10.2	10.1	11.1	5.2	–3.2	–0.7	–0.7	–5.06	–2.9	na	na
c) Trade dues and other current liabilities	17.9	8.2	12.8	12.8	17.2	20.2	14.3	14.3	20.3	18.5	17.0	17.9
C. Total (A+B)	100	100	100	100	100	100	100	100	100	100	100	100

Source: Securities and Exchange Board of India (2008).

note that with the onset of the reforms and the liberalization of the capital markets, the external sources of funds have actually come down till 2004–5 and in the last two years external financing has, once again, become important but with debt capital becoming more important. It should of course be mentioned that the data for the last two years 2005–6 and 2006–7 need to be taken with some caution as the number of companies covered by the Central Bank of India (RBI) survey, from which these numbers have been derived, shows a dramatic halving compared to those covered in the previous years. So it may well be possible that the increased share of external finances is actually a statistical artefact. Companies seem to be depending increasingly on self-generation. Within the external source of finance, bank borrowings are more important (due to the current global financial crisis and with the likely existence of a liquidity crunch within the banking system, bank borrowings, despite the steps taken by RBI, can become very tight). More recent data from RBI (contained in RBI 2008a) also confirm this trend. While this pattern of financing, with the internal generation accounting for the larger share, may be important for existing companies, new companies may have to depend on external sources. For this the emergence and growth of the private equity market and the venture capital funding has been somewhat helpful.

9.5.2.1 EVIDENCE FROM NEN MICRODATA

The microdata further complement our earlier finding that as of now most companies place a heavy reliance on internal sources of finance. This is all the more evident in the NEN start-up dataset, where we observe that over 70 per cent of the entrepreneurs (Table 9.13) relied on personal savings for their initial funding needs and about 40 per cent of them continue to rely on the same personal sources for their additional funding.

Table 9.13 Sources of initial and additional funding of the National Entrepreneurship Network start-ups

Primary source of initial funding			Primary source of additional funding		
Source	Total	%	Source	Total	%
Angel investors	21	4	Angel investors	66	11
Bank loans	22	4	Bank loans	45	8
Family and friends	102	17	Family and friends	113	19
Personal savings	409	70	Personal savings	234	40
Personal credit cards	4	1	Personal credit cards	12	2
Venture capitalists	9	2	Venture capitalists	48	8
Not mentioned	21	4	Not mentioned	70	12
Total	588	102	Total	588	100

Source: National Entrepreneurship Network Online <http://www.nenonline.org/>.

However, we see that external sources such as bank loans, venture capital, and angel investors become very important for expansion of the scale of activity. It is interesting to note that equity capital is conspicuous by its absence as a source of funding even at the expansion stage. These additional sources of venture capital and angel investments are analysed in some depth below. In fact we argue that both venture capital and angel investing are market-based solutions to market failures in the financing of knowledge-intensive entrepreneurship.

9.5.2.2 GROWTH OF PRIVATE EQUITY AND VENTURE CAPITAL IN INDIA

Universally private equity (PE) and venture capital (VC) have been the main source of risk capital for technology-based entrepreneurs. But there are some differences between these two, namely that VC focuses on investing in private, young, fast growing companies. Buyout and mezzanine investing focuses on investing in mature companies. The history of the VC industry in India can be traced to the late 1980s (Mani 1997) and since then the history of the fledgling industry can be divided into four phases (Indian Venture Capital Association 2007):

- Phase I: Formation of Technology Development and Information Company of India, TDICI in the 1980s and regional funds as Gujarat Venture Fund Limited, GVFL, and Andhra Pradesh Industrial Development Corporation, APIDC in the early 1990s
- Phase II: Entry of foreign VC funds between 1995 and 1999
- Phase III: (2000–7). Emergence of successful India-centric VC firms
- Phase IV: (2007) Global VCs and PE firms actively investing in India

At this point it is necessary to point out that there are no official sources of data on VC in the country, but what is available in the public domain is from the website of the Indian Venture Capital Association and it clubs both VC and PE deals together (Table 9.14). However, we have obtained the share VC in

Table 9.14 Growth of the private equity/venture capital industry in India, 2000–7

	Number of deals	Value of deals	Average per deal
2000	280	1,160	4.14
2001	110	937	8.52
2002	78	591	7.58
2003	56	470	8.39
2004	71	1,650	23.24
2005	146	2,200	15.07
2006	299 (92)	7,500 (508)	25.08 (5.52)
2007	387 (98)*	14,234 (543)*	36.78 (5.54)

Note: * Figures in brackets are the VC deals.

Source: Indian Venture Capital Association (2007); US-IVCA/Venture Intelligence (2006 and 2007).

the total PE from another reliable private source of data (US-IVCA/Venture Intelligence 2006 and 2007). The phenomenal growth of the PE/VC industry can be gauged from the fact that the average size of a deal has shown an increase of 51 per cent per annum since 2000. However, based on the data provided in US-IVCA/Venture Intelligence (2006 and 2007) real VC investments in 2007 were only 4 per cent in terms of total value of deals, but about 25 in terms of the number of deals.

About two-thirds of the value of deals have gone towards the IT and Information Technology Enabled Services industry. Although the VC industry is largely private- and foreign-owned, the government has played a very important role in establishing the industry and nurturing it through a variety of fiscal concessions (Mani 1997). Once again, the growth of the VC industry has provided some financial support to knowledge-intensive entrepreneurship and is thus a market-based solution to a market failure in the financing of knowledge-based entrepreneurship.

9.5.2.3 GROWTH OF ANGEL FUNDING

Entrepreneurs who have untested business models or innovative ideas typically get their first round of funding from angel investors. If and when their business model works and they are ready for scale up, they approach venture capitalists who usually invest more money (at least Rs. 250 million) in the company in return for an equity stake. Angel investors broadly differ from venture capitalists in the scale of funding. Besides, angel investors invest their personal wealth as opposed to venture capitalists who mostly work as fund managers. The size of the angel investments have been variously estimated to about Rs. 10 billion in 2007. There is an inexorable link between the growth of angel investment and the growth of High Networth Individuals (HNIs).[13] This can be further explained as follows.

Though the risk with start-ups is much higher than other asset classes such as real estate, equity, mutual funds, commodities, and sometimes even art funds, HNIs are betting on the opportunity of considerably higher returns associated with start-ups. To institutionalize this process of channelling funding from HNIs to technology-oriented start-ups, the Indian Angel Network[14] was founded in 2006. Around 80 HNIs are part of this network today. In the recent past, the angel community has grown considerably in India. A typical investment by an HNI in a start-up falls in the range of Rs. 1 to 5 million and the exit duration is usually between four to seven years. The returns, on the other hand, can vary from 400 per cent to even zero if the investment goes bad. Hitherto the network has supported around 12 technology-oriented ventures primarily in the arena of IT software.

In a bid to promote funding for start-ups, the government plans to offer tax breaks to angel investors, who provide a part of their personal wealth as seed capital for such firms. A proposed legislation, the National Innovation Act,[15] envisages doing away with the stamp duty currently levied on shares held by angel investors and the tax imposed on profits they make in early stage firms. However, these tax breaks would apply only to companies that are incubated in designated areas—called special innovation zones[16]—and are likely to include technology parks and incubation facilities of academic institutions such as the Indian Institutes of Technology (IITs).

9.5.3 *Government and public–private partnership programmes*

There are a number of government programmes and institutional arrangements that are put in place to encourage technology-based entrepreneurship mostly by the central government but in some cases by individual state governments as well. One of the earlier attempts has been the establishment of the National Science and Technology Entrepreneurship Development Board (NSTEDB) in 1982 under the administrative purview of the Department of Science and Technology. It is an institutional mechanism, with a broad objective of promoting gainful self-employment amongst the science and technology manpower in the country and to set up knowledge-based and innovation-driven enterprises. The NSTEDB has two major responsibilities: the first is to establish technology parks and incubators for nurturing already existing entrepreneurs. It thus provides the crucial infrastructure and other value added support for growing entrepreneurs. Second, it organizes a series of training programmes to initiate freshly graduated engineers and other technically qualified students to learn the nitty gritty of entrepreneurship. The actual achievements of NSTEDB in carrying out these two responsibilities are outlined in Appendix table 9.1.

Another important programme to facilitate knowledge-intensive entrepreneurship is the Technopreneur Promotion Programme (TePP) administered by the Department of Scientific and Industrial Research of the Government of India. The programme was launched in 1998 to help realize the vast latent innovative potential of the people. The basic objective of TePP is for individual innovators to emerge as technopreneurs—technology-oriented entrepreneurs. TePP's support is provided for in all areas except software development for which there are other avenues of support. It helps the inventor to identify and network with an appropriate R&D/academic institution for guidance, technical consultancy, and development of models or prototypes. It assists in filing for and securing intellectual property rights and last but not the least, in linking with appropriate sources of finances for commercialization of the product. TePP by itself provides financial support of up to Rs.1 million as a grant-in-aid to prove the idea and a similar amount for the second phase for commercialization. As of 31 March 2008, about 80 projects are under various stages of completion. But there is precious little information on the number of entrepreneurs that may have emerged consequent to this programme.

9.5.3.1 PUBLIC–PRIVATE PARTNERSHIP FOR REDUCING INFORMATION ASYMMETRIES

Technology Innovation Management and Entrepreneurship Information Service (TIME IS), a joint project of NSTEDB, the Department of Science and Technology (DST), and the Federation of Indian Chamber of Commerce and Industry (FICCI), is now one of the credible ladders towards the enhancement of India's entrepreneurial economy. The project has taken initiatives to provide guidance and assistance to the entrepreneurs especially the technopreneurs to find technologies, projects, funding options, and information about policy environment, incentive schemes, and industrial infrastructure available in the country covering both the central and state government and have become proficient at tapping the local talent pool. TIME IS facilitates entrepreneurs with 'online interactive tools and templates' for developing 'project profile', 'feasibility reports', calculating 'financial and profitability ratios', and estimating the 'market potential'.

9.5.3.2 THE DST-LOCKHEED MARTIN INDIA INNOVATION GROWTH PROGRAMME

This programme is a two-year, nationwide project, created to enhance the growth and development of India's entrepreneurial economy. The programme is wholly funded by the Lockheed Martin Corporation, and was developed with the assistance of the IC2 Institute at the University of Texas, and FICCI. Its overall goal is to accelerate the launch of Indian early stage technologies into the global marketplace. The programme features a competitive selection process. Selected participants may receive specialized training and funding opportunities. Top selectees will also be eligible to receive professional business development support to assist them in entering global markets.

9.5.3.3 BUSINESS INCUBATORS IN INDIA[17]

This is a relatively new concept in India and unlike in other BRIC countries such as Brazil or China, even now the concept is not so well-developed in India as a support system for engineering technology-based entrepreneurs. As of 2007, there are approximately 40 incubators spread throughout the country.[18]

There are essentially three types of incubators depending upon the physical location and ownership of these incubators. They are those which are established: (i) under the aegis of leading institutions of engineering technology and management, for instance within the IIMs and the IITs, (ii) within the Science and Technology Entrepreneurs Parks (STEPS), for instance within the Technopark in Trivandrum, and (iii) by leading private sector enterprises such as Nirmalabs.

The history[19] of business incubators in the country could be traced to 1985 when the Tiruchirappalli Regional Engineering College—Science and Technology Entrepreneurs Park (TREC—STEP) was set up within the Regional Engineering College at Tiruchi (now known as National Institute of Technology). One can see two broad phases in the growth of incubators in the country: the first phase is from 1986 through the late 1990s when incubators were set up in broad based STEPS and the second phase from the late 1990s to the present when more focused Technology Business Incubators (TBI) were established within the leading national institutes of technology, engineering, and management.

The incubation idea has received a fillip with the NSTEDB deciding to create an incubation fund with an initial corpus of Rs. 50 million to facilitate the development of entrepreneurship in knowledge-based, high-growth businesses. The corpus of Rs. 50 million would be allocated in equal measure to five out of the 32 STEPs and TBIs that were under the administrative ambit of the Centre's Department of Science & Technology. The idea behind the initiative was to ensure that technology entrepreneurship based on business ideas was not hindered for want of initial funds required for market research, etc. The fund is essential for bridging the financing needs of a technology-based entrepreneur between the time he floats his business venture and the time it begins to attract venture funding.

Since the incubator idea itself is new and evolving, there are no detailed studies on the effectiveness of these as an instrument for promoting technology-based entrepreneurship except that it is an experiment which is worth watching especially when a number of prominent centres such as the IIM-Ahmedabad are engaged in the nurturing of this fledgling idea of incubation and entrepreneurship among its student community.

9.5.4 *Private sector initiatives in promoting entrepreneurship*

The 1990s external environment is characterized by the importance it has attached to enlarging the scope of market forces in all matters with respect to economy. Since the creation of private entrepreneurship and its sustenance is a necessary condition for increasing the share of the private sector in India's economy, one sees an increased activity by various private sector agencies towards creating this activity on their own in addition to complementing the efforts of the state through various public–private initiatives. Two such private sector initiatives that have become very active in recent times are: (a) the TiE network (The Indus Entrepreneurs),[20] and (b) the Wadhwani Foundation, which seek to promote entrepreneurship by organizing workshops and seminars nationally. Founded by entrepreneur Romesh Wadhwani, the foundation funds various entrepreneurship education related projects like NEN,[21] which brings together prestigious Indian higher education institutions and entrepreneurs. Again both these private initiatives are more mentoring in nature and act as a 'catalyst' for the creation of technology-based entrepreneurship. Of the two, NEN has been particularly active in catalyzing technology-based entrepreneurship. Of the various strategies that it has employed is the NEN hottest start-up competition.

9.5.5 *Education and training*

The general assumption is that the country has a veritable supply of scientists and engineers and many of these who graduate from a variety of universities and technological institutes can be trained and encouraged to become entrepreneurs. This assumption is usually substantiated by invoking the gross enrolment of the undergraduate degree programmes in engineering and architecture: the enrolment has increased from 7.08 lakhs in 2002–3 to 16.68 lakhs in 2005–6 (Ministry of Higher Education 2008). Given high failure and drop out rates, these enrolments need not translate directly into a quantum jump in graduation rates, but they do indicate the likely addition to the stock of technical manpower in the country. Questions have often been raised regarding the quality of these graduates which varies highly. We had also seen (from the NEN start-up dataset) that most of the successful entrepreneurs have either studied in Tier I institutes such as the IITs or NITs or they have secured degrees from foreign universities.[22] This is because a typical engineering graduate does not receive much training in becoming an entrepreneur during their four year degree programme. Although most engineering graduate level programmes have management and economics papers in their curriculum these are in most cases badly taught and largely theoretical in nature.

9.5.5.1 MORE FORMAL TRAINING IN ENTREPRENEURSHIP

In order to give formal training to engineering graduates in entrepreneurship training, as mentioned earlier, the NSTEDB has actually a limited number of schemes in this direction. Even an exclusive national institute devoted to entrepreneurship training in the form of the Entrepreneurship Development Institute has been established at Ahmedabad for this purpose. The NEN and Indian Institute of Management (IIM) at Bangalore have also initiated a number of courses of varying duration and content to provide systematic training in this area.[23]

Finally it may be said that the positive entrepreneurial ecosystem (consisting of market opportunities, government and private support systems of various kinds, VC, etc.) is beginning to have a positive impact on the likelihood of students graduating from Tier I institutes taking up entrepreneurship as a career compared to the lure of the labour market employment (Bansal 2008; Basant and Chandra 2008).

9.6 Conclusions

Our study has sought to highlight a number of positive indicators as far as technology-based entrepreneurship in India is concerned in the post-liberalized regime. Following Gupta (2001), the state has to promote four facilitating factors for technology-based entrepreneurship to be sustained. They are: (1) creating the right environment for success: entrepreneurs should find it easy to start a business; (2) ensuring that entrepreneurs have access to the right skills. According to Gupta (2001) most Indian start-up businesses face two skill gaps: entrepreneurial (how to manage business risks, build a team, identify and get funding) and functional (product development know-how, marketing skills, etc.). In other countries, entrepreneurs either gain these skills by hiring managers or have access to 'support systems' such as universities or other institutions that may nurture many regional businesses. In addition, business schools give young graduates the skills and knowledge required for business today. India can move toward ensuring that the curriculum at universities is modified to address today's changing business landscape, particularly in emerging markets, and to build 'centres of entrepreneurial excellence' in institutes that will actively assist entrepreneurs; (3) ensuring that entrepreneurs have access to 'risk' capital: for a long time, Indian entrepreneurs have had little access to capital. As mentioned earlier it is a fact that in the last few years, several VFs have entered the Indian market; and (4) enabling networking and exchange: entrepreneurs learn from experience—theirs and that of others. Much of the success of Indians in Silicon Valley is attributed to the experience, sharing, and support which members of organizations such as TiE have extended to young entrepreneurs. Given the positive contribution of knowledge-intensive entrepreneurship and given the ongoing financial crisis which would turn some of the facilitating factors into strong constraints, one should not underestimate the catalytic role that the government has to play in supporting the growth of these desirable activities.

APPENDIX

Appendix table 9.1 Achievements of the NSTEDB (2007)

Programme	(Physical Achievements)
Entrepreneurship Awareness Camp (EAC)	
Conducted	1,850
Students exposed	110,000
Entrepreneurship Development Programme (EDP)	
Conducted	717
Persons trained	16,159
Technology-based EDPs (TEDP)	
Conducted	413
Persons trained	8,450
Faculty Development Programme (FDP)	
Conducted	160
Faculty trained	3,200
Entrepreneurship Development Cell (EDC)	
ED Cells established	55
ED Cells being supported currently	36
S&T entrepreneurs park (STEP)	
Number of STEPs	15
Units set up	910
Jobs generated	6,300
Technology Business Incubator (TBI)	
Number of TBIs	15
Units set up	85
S&T entrepreneurship development (STED) Project	
STED Projects currently operating	42
Skill Development through S&T (STST)	
Persons trained	113,000

Notes

1. The fact that the break in the trend growth rate of India's GDP has occurred in 1991 has sparked off a lively debate with some analysts holding the view that this occurred earlier in the 1980s. However after examining the various issues, technical and otherwise, the consensus is for the break to have occurred in 1991 itself. For a succinct summary of this debate see Basu (2008).
2. *Research Policy*, 32 (2): 181–35.
3. For a concise summary of this growth performance, see Basu (2008) and Panagariya (2008).
4. Even though the largest conglomerate group in the country, the Tata group, has expanded massively during this period (its total sales revenue now accounts for over 3 per cent of India's GDP), much of its growth has been outside the country. For instance, according to the group's website <http://www.tata.com/tataworldwide/index.aspx?sectid=1y2Y3CZ5A2s=>, in 2007–8 about 61 per cent of its total sales were derived from its international operations.

5. There is now a small but growing literature on the growth and emergence of Indian multinational corporations. See, for instance, Nayyar (2008).
6. The Global Entrepreneurship Monitor (GEM) research programme is an annual assessment of the national level of entrepreneurial activity. Started as a partnership between the London Business School and Babson College, it was initiated in 1999 with ten countries, expanded to 21 in the year 2000, with 29 countries in 2001 and 37 countries in 2002. GEM 2007 conducted research in 42 countries. The research programme, based on a harmonized assessment of the level of national entrepreneurial activity for all participating countries, involves exploration of the role of entrepreneurship in national economic growth.
7. GEM considers five measures of entrepreneurial activity: (i) nascent entrepreneurship rate, which is the percentage of the 18–64-year-old population who are currently a nascent entrepreneur, i.e. actively involved in setting up a business they will own or co-own; this business has not paid salaries, wages, or any other payments to the owners for more than 3 months; (ii) new business ownership rate, the percentage of the 18–64-year-old population who are currently an owner/manager of a new business, i.e. owning and managing a running business that has paid salaries, wages, or any other payments to the owners for more than three months, but not more than 42 months; (iii) early stage entrepreneurial activity, the percentage of the 18–64-year-old population who are either a nascent entrepreneur or owner/manager of a new business (as defined above); (iv) established business ownership rate, the percentage of the 18–64-year-old population who are currently an owner-manager of an established business, i.e. owning and managing a running business that has paid salaries, wages, or any other payments to the owners for more than 42 months; and (v) overall entrepreneurial activity rate, that is the percentage of the 18–64-year-old population who are either involved in early stage entrepreneurial activity or owner/managers of an established business (as defined above).
8. NEN, founded in 2002, is a not-for-profit initiative of the Wadhwani Foundation, working to inspire, educate, and support the next generation of high-growth entrepreneurs in India. NEN was cofounded by five of India's premier academic institutions: IIT Bombay, IIM Ahmedabad, SP Jain Institute Bombay, IBAB Bangalore, and BITS Pilani. Over the past three years, NEN's focus on introducing a new paradigm in entrepreneurship education in India—and its innovative method of doing so—has made it its leading catalyst on campuses across India. NEN's ultimate goal is to help launch thousands of new entrepreneurs, who in turn will create hundreds of thousands of much needed valuable jobs for India.
9. To be a technology company, the following three sets of conditions must be fulfilled, namely: (i) it owns proprietary technology that contributes to a significant portion of the company's operating revenues, (ii) it manufactures a technology-related product, and/or (iii) it devotes a significant proportion of operating revenues to research and development.
10. For details, see <http://www.deloitte.com/dtt/article/0,1002,cid%253D206929,00.html> (accessed on 10 December 2008).
11. See GEM (2007).

12. See also National Knowledge Commission (2008).
13. HNIs hold at least US$1 million in financial assets, excluding collectibles, consumables, consumer durables, and primary residences. According to the World Wealth Report 2008 prepared by Capgemini and Merril Lynch the number of HNIs in India has gone up by 23 per cent in 2007 compared to 2006 and there are about 123,000 HNIs in India as of 2007. Further, the report said that the combined wealth of the HNIs has increased to US$440 billion until 2007. The rapid expansion of economy, increased foreign investment, increase in the savings rates, and gains on the country's stock markets are the prime factors responsible for increase in the number of Indian HNIs. As of December 2007, HNIs in India have investible surplus of more than US$1 million.
14. The Indian Angel Network is India's first and largest Angel network with successful entrepreneurs and high profile CEOs interested in investing in early stage businesses across India, which have potential to create disproportionate value. The Network has invested in multiple sectors like IT, intellectual property, hospitality, mobile, education, Internet, etc.
15. For details of the draft National Innovation Act, see the website of the Government of India's Department of Science and Technology at <http://dst.gov.in/draftinnovationlaw.pdf> (accessed on 7 December 2008).
16. Although the government is yet to notify these so-called SIZs, the recently established biotechnology cluster at Mohali in Punjab, the IIT Madras Research Park, etc., will qualify for this status.
17. For a systematic survey of the role of incubators in facilitating knowledge-intensive entrepreneurship, see Akçomak (Chapter 10 in this volume).
18. The number of incubators in India compares very poorly with that of China: China had 500 incubators in 2004 employing six-lakh people as against a handful in India. The incubation market place in China, though relatively recent, is well-developed, with the government playing a predominant role in the business of incubation by channeling resources to tie up with the mandate of high-technology-led economic growth.
19. This is based on a presentation by R. M. P. Jawahar, Executive Director of Triuchirappalli Regional Engineering College—Science and Technology Entrepreneurs Park (TREC—STEP), available at <http://www.infodev.org/en/Publication.34.html>.
20. TiE was founded in 1992 in Silicon Valley. It is a network of entrepreneurs, professionals, and venture capitalists active worldwide in technology-related sectors, who share the same geographical and cultural origins. At present TiE can count on over 10,000 members sub-divided into 44 chapters in nine countries including among others the USA, Australia, UK, Singapore, United Arab Emirates, and India. Every year TiE holds a conference in Silicon Valley attended by numerous stakeholders from the IT industry. Although TiE is not directly involved in funding enterprises, it may provide important mentoring services to its affiliates and would-be entrepreneurs in many ways, such as assistance in preparation of business plans, fundraising, or strategic guidance.
21. NEN is working to inspire, educate, and support the younger generation of entrepreneurs in India. NEN helps its 403 member academic institutes build

comprehensive, high impact entrepreneurship programmes on their campuses. It reaches over 400,000 young people, has helped launch more than 350 student entrepreneurship clubs, and has an individual membership base of more than 60,000.

In addition to working with institutes, NEN provides support to India's growing pool of young and future entrepreneurs. In 2008 NEN also launched the TATA-NEN hottest start-ups awards, a national programme to increase the visibility of high potential start-ups, and engage the public to create more support for these start-ups.

22. In order to improve the quality of technical education, the central government is in the process of establishing a number of Tier I institutions of higher learning in engineering and management. In addition to 30 central universities during the 11th Plan period (2007–12), the government will also set up eight IITs, ten National Institutes of Technology, 20 Indian Institutes of Information Technology as far as possible in the public–private partnership mode, three Indian Institutes of Science Education and Research, seven Indian Institutes of Management (IIMs), and two Schools of Planning and Architecture.
23. In 2005, 16 colleges across the country sent faculty members to the entrepreneurship educators course conducted by NEN; at the end of 2007, 269 colleges across the country had signed on to the programme—an initiative of the Wadhwani Foundation, which is focused on accelerating entrepreneurship in emerging economies, and has trained more than 470 faculty members in Indian colleges so far. The N. S. Raghavan Centre of Entrepreneurial Learning in the Indian Institute of Management, Bangalore, in collaboration with the Singapore-based Universitas 21 Global, a distance learning educator, has started a programme for entrepreneurial training in family-owned businesses. Both N. S. Raghavan Centre of Entrepreneurial Learning and the Stanford Technology Ventures Programme, the entrepreneurship centre at Stanford University's School of Engineering, provide faculty and learning material for the NEN programme, which trains teachers who can in turn teach entrepreneurship as an academic course to college students across India.

References

Bansal, R. (2008) *Stay Hungry, Stay Foolish.* Ahmedabad: Indian Institute of Management.

Basant, R., and P. Chandra (2008) 'Role of Educational and R&D Institutions in City Clusters: An Exploratory Study of Bangalore and Pune Regions in India', *World Development*, 35 (6): 1037–55.

Basu, K. (2008) 'The Enigma of India's Arrival: A Review of Arvind Virmani's, *Propelling India: From Socialist Stagnation to Global Power*', *Journal of Economic Literature*, 46 (2): 396–406.

Central Statistical Organization (CSO) (2008) 'National Accounts Statistics 2008'. New Delhi: CSO.

Damodaran, H. (2008) *India's New Capitalists, Caste, Business, and Industry in a Modern Nation*. New Delhi: Palgrave Macmillan.

Department of Science and Technology (2006) 'R&D Statistics'. New Delhi: Government of India, Department of Science and Technology.

Evaluserve (2008) 'Patenting Landscape in India', White Paper. New Delhi: Evaluserve.

Global Entrepreneurship Monitor (GEM) (2007). '2007 Executive Report'. Available at: <http://www.gemconsortium.org/about.aspx?page=pub_gem_global_reports>.

Gupta, R. (2001) 'Creating Indian Entrepreneurs', *India Today*, 12 February.

Indian Venture Capital Association (2007) 'Venture Capital and Private Equity in India'. Presentation slides available at <http://www.indiavca.org>.

Karki, R. (2008) *Competing with the Best*. New Delhi: Penguin Books Private Limited.

Khanna, T. (2008) *Billions of Entrepreneurs: How China and India Are Reshaping Their Futures—and Yours*. New Delhi: Penguin Books India.

KPMG-TiE (2008) 'Entrepreneurial India—An Assessment of the Indian Entrepreneur's Confidence in their Business Ecosystem'. Available at: <http://www.in.kpmg.com/TL_Files/Pictures//TiE08.pdf>.

Kroeber, A., and R. Yao (2008) 'Large and in Charge', *Financial Times*, 14 July.

Mani, S. (1997) 'Financial Barriers to Domestic Technology Development and Venture Capital', *Science, Technology and Development*, 15 (2 and 3): 227–40.

——— (2007) 'Government Support for Sustaining a Knowledge-based economy—An Examination of India's Technology Policy Initiatives, 1990–2005'. In A. Vaidyanathan and K. L. Krishna (eds), *Institutions and Markets in India's Development, Essays for K. N. Raj*. New Delhi: Oxford University Press.

Ministry of Company Affairs (2007) '50th Annual Report for the Year ended March 2006'. New Delhi: Government of India, Ministry of Company Affairs.

Ministry of Higher Education (2008) 'Annual Report 2007–08'. New Delhi: Government of India.

Mohan, R. (2006) 'Economic Reforms and Corporate Performance in India'. Special Talk at the FICCI-IBA Conference on Global Banking: Paradigm Shift Towards Meeting the Emerging Challenges. Mumbai, India, 26–8 September.

National Knowledge Commission (2008) 'Entrepreneurship in India'. New Delhi: National Knowledge Commission.

Nayyar, D. (2008) 'The Internationalization of Firms from India: Investment, Mergers and Acquisitions', *Oxford Development Studies*, 36 (1): 111–31.

NSTEDB (2007) 'Annual Report 2006–07'. New Delhi: Government of India.

Panagariya, A. (2008) *India: The Emerging Giant*. New York: Oxford University Press.

Reserve Bank of India (RBI) (various issues). 'Annual Report'. Mumbai: Reserve Bank of India.

——— (2006) 'Remittances from Overseas Indians: A Study of Methods of Transmission, Cost and Time', *Reserve Bank of India Bulletin*, November: 1361–69.

——— (2008a) 'Annual Report 2007–08'. Mumbai: Reserve Bank of India.

——— (2008b) 'Invisibles in India's Balance of Payments: An Analysis of Trade in Services, Remittances and Income', *Reserve Bank of India Bulletin*, February: 261–302.

Secretariat of Industrial Assistance (2008) 'SIA Statistics' (August). Available at <http://siadipp.nic.in/publicat/stats/aug2008/index.htm>.

Securities and Exchange Board of India (2008) 'Handbook of Statistics of the Indian Securities Market 2008'. Mumbai: Securities and Exchange Board of India.

Taube, F. (2009) 'Diversity and the Geography of Technology Entrepreneurship: Evidence from the Indian IT Industry'. In M. Keilbach, J. Pawan Tamvada, and D. B. Audretsch (eds), *Sustaining Entrepreneurship and Economic Growth–Lessons in Policy and Industry Innovations from Germany and India*. Springer: New York.

US-IVCA/Venture Intelligence (2006 and 2007). 'India Venture Capital Report 2006 and 2007'. Chennai: Venture Intelligence.

10

Incubators as Tools for Entrepreneurship Promotion in Developing Countries

İ. Semih Akçomak

10.1 Introduction

Researchers and industry practitioners highlight the role that small and medium-sized enterprises (SMEs) play in creating income and employment. Because of their flexible structure SMEs quickly adapt to changes in the economic environment and technology and as such small firms are the cornerstone in policy-making regarding new venture and job creation (Wennekers and Thurik 1999). For instance, Chinese small firms account for about 90 per cent of all firms and about 67 per cent of all firms with any science and technology activity (Lundin, Sjöholm, and Qian 2006), micro-enterprises (less than ten employees) make up 96 per cent of all firms and employ over 45 per cent of the economically active population in Peru (Jäckle and Li 2006). However, small and new ventures have several disadvantages that most entrepreneurs face. They cannot benefit from scale economies both from the output and input side. Small size is an important constraint for process and product innovations. Moreover, difficulties in gaining access to tangible and intangible resources, poor management skills, and lack of know-how hamper survival rates among (high-tech) new ventures (Allen and Rahman 1985; Smilor and Gill 1986; Peters, Rice, and Sundararajan 2004; Gassman and Becker 2006).[1] These drawbacks that are common to entrepreneurs in developed countries are exacerbated in developing countries due to additional impeding factors, such as lack of human capital, high macroeconomic volatility, and poor functioning formal institutions.

Incubators provide an attractive framework to practitioners in dealing with the difficulties in the process of entrepreneurship summarized above. They can be considered as a remedy for the disadvantages that small and new firms encounter by providing numerous business support services and they are useful in fostering technological innovation and industrial renewal (Similor and Gill 1986; Allen and McCluskey 1990; Mian 1996a). They can be viewed as a mechanism, (i) to support regional development through job creation (Allen and Levine 1986; Mian 1997; Roper 1999; Thierstein and Wilhelm 2001), (ii) for new high-tech venture creation, commercialization, and transfer of technology (Mian 1994, 1997; Phillips 2002; McAdam and McAdam 2008), and (iii) an initiative to deal with market failures relating to knowledge and other inputs of innovative process (Colombo and Delmastro 2002). Studies have shown that one-third of new firms do not survive the third year and about 60 per cent do not survive the seventh year (Coris 2007). This number drops to 15–20 per cent among incubator tenants (Bruton 1998; Adegbite 2001; Lalkaka 2002; Abetti 2004). For these reasons many countries have increasingly been engaged in establishing incubators. It is estimated that there are around 3,500 incubators worldwide: one-third are located in North America, about 30 per cent in Europe, and the remaining 40 per cent in developing countries of which most are in Brazil and China.

This chapter's purpose is to discuss the rationale for the adoption of incubators in developing countries. As such the discussion elaborates questions like, what particular benefits can incubators provide for entrepreneurs and small companies? What factors (resources) are important in a successful incubator model? Can developing countries sufficiently harness tangible and intangible resources for the success of incubators?

10.2 Review of the incubator literature

The literature on incubators can be classified into two broad categories. The first set of studies deals with the theory of the incubators and the incubator model and investigates how incubators are formed, what their aims are, how they are planned, and how they are managed (e.g. Similor and Gill 1986; Allen and McCluskey 1990; Aernoudt 2004; Leblebici and Shah 2004; Grimaldi and Grandi 2005; Becker and Gassmann 2006). The second set of studies evaluates incubators regarding certain factors that define success indicators. These papers mainly focus on whether incubators have achieved their goals in supporting entrepreneurs, encouraging creation of new firms and jobs, and establishing an entrepreneurial society (e.g. Mian 1996a; Colombo and Delmastro 2002; Peters, Rice, and Sundararajan 2004; Rothaermel and Thursby 2005a, 2005b; Aerts, Matthyssens, and Vandenbempt 2007).

10.2.1 *Brief history and typology of incubators*

We focus on the concept of (technology) incubators in this chapter. There are many derivatives that spun off from the concept of science parks and incubators. The International Association of Science Parks defines a science park as: 'an organization managed by specialized professionals, whose main aim is to increase the wealth of its community by promoting the culture of innovation and the competitiveness of its associated businesses and knowledge-based institutions'. To enable these goals, a science park: (i) stimulates and manages the flow of knowledge and technology amongst universities, R&D institutions, companies, and markets; (ii) facilitates the creation and growth of innovation-based companies through incubation and spin-off processes; and (iii) provides other value added services together with high quality space and facilities. For instance, a University Research Park is a derivative of this concept. The main differences are the contractual and/or operational relationship with a university and the focus on the transfer of technology and promotion of R&D under university–industry partnership. On the other hand the National Business Incubation Association (NBIA) defines business incubation as

> a dynamic process of business enterprise development which: (i) nurtures young firms that is, helps them to survive and grow during the start-up period when they are most vulnerable; (ii) provides hands-on management assistance, access to financing and orchestrated exposure to critical business or technical support services; and (iii) offers entrepreneurial firms shared office services, access to equipment, flexible leases, and expandable space—all under one roof.

The most important element that identifies incubators is that they provide high level business support/management services under one roof for entrepreneurs and new ventures that have (medium) high level technological focus to create synergy (e.g. Allen and McCluskey 1990; Aerts, Matthyssens, and Vandenbempt 2007).[2]

As stated before, this chapter covers the literature on incubators. We do not provide detailed definitions of other forms, such as technoparks, industrial parks, technopolis, etc. All these establishments share similar characteristics and vary on relatively minor points. The differences between various definitions and forms could easily be analysed with the help of Figure 10.1 where different derivatives are placed on a continuum from low to high management support services and from low to high technological level. Most contemporary incubators originate from multipurpose business incubators and business centres that were established in the 1970s. Today's incubators have highly selective admission criteria, and provide hands-on business and management assistance for new ventures that are higher in technological continuum and

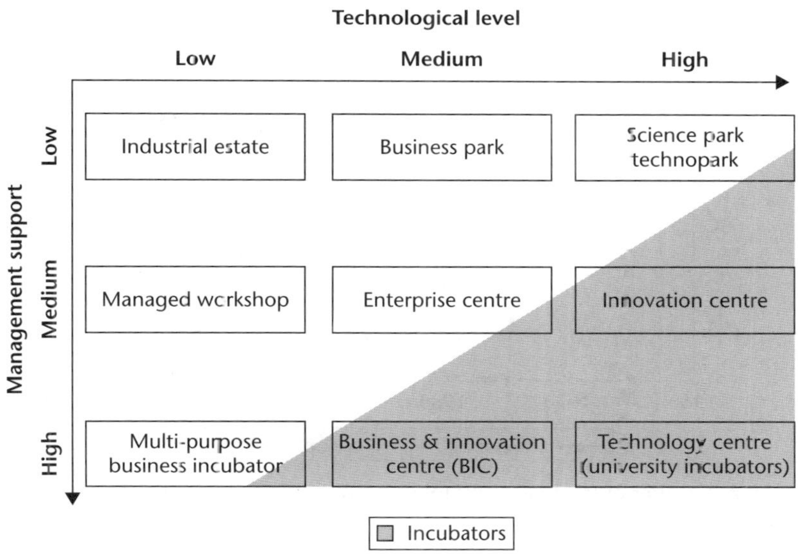

Figure 10.1 Identifying incubators among other forms of similar establishments
Source: Centre for Strategy and Evaluation Services (2002).

have a high potential to create revenue and jobs. The incubator, as understood in the current terminology, is represented in the grey shaded area in Figure 10.1.

10.2.1.1 DEVELOPMENT OF THE INCUBATION IDEA

It is widely accepted that the first incubator was created by Joseph Mancuso in Batavia, New York in 1957 on a former Massey–Ferguson facility (Leblebici and Shah 2004; O'Neal 2005). The incubator movement was initiated by the managers of the incubators in the US through a series of conferences in the mid-1980s sponsored by the US Small Business Administration (Allen and McCluskey 1990). Incubators differed from the existing industrial parks and estates as the focus was value added business services rather than real estate development and subsidized rents (O'Neal 2005). These developments finally led to the foundation of NBIA in 1985, with 40 founding members. Figure 10.2 depicts the number of incubators established in the developed and developing world in five-year intervals starting from the 1960s.

One can categorize incubators either according to their funding scheme (sponsors) (e.g. Becker and Gassmann 2006) or according to their main objectives (e.g. Aernoudt 2004). Despite numerous similar typologies, there is still no theoretical study that aims at merging these two seemingly separate typologies. This chapter makes an attempt in this direction. The basic framework is presented in Figure 10.3. The vertical dimension locates incubators on a

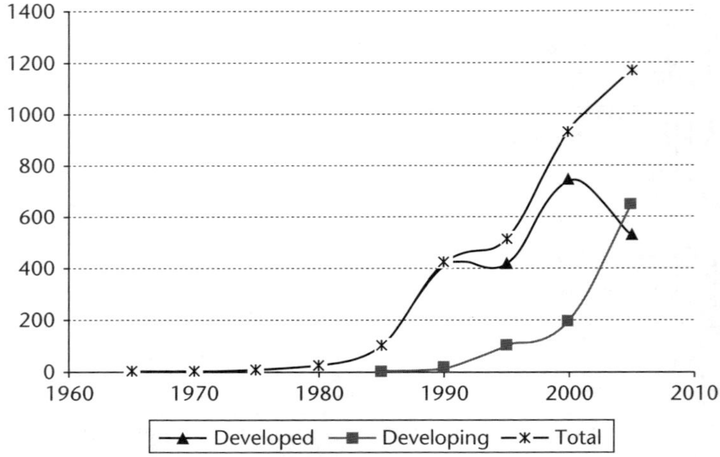

Figure 10.2 Number of incubators established in developed and developing countries
Source: Author's own calculations.

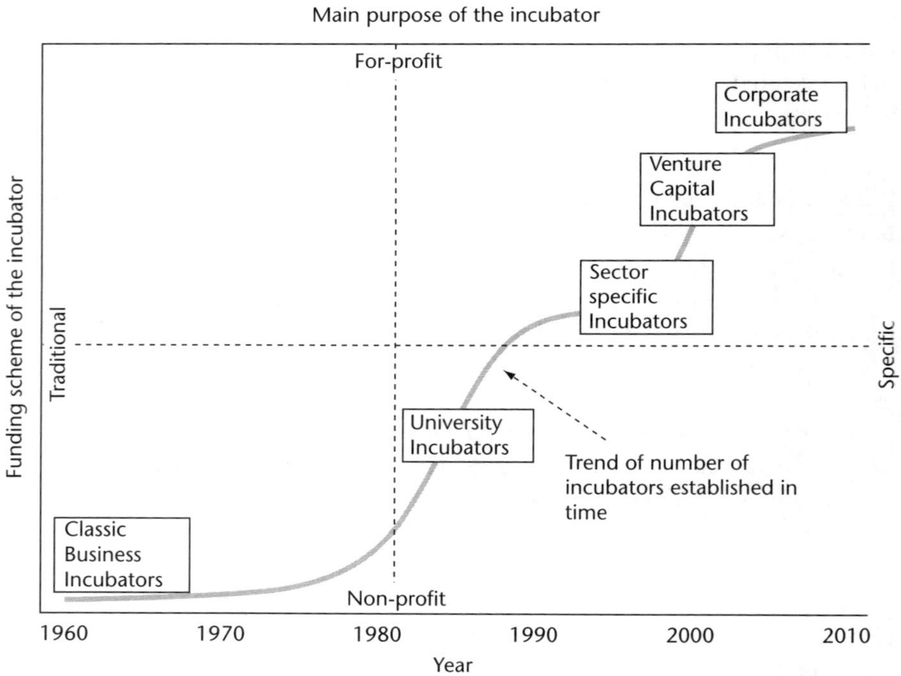

Figure 10.3 Typology and the evolution of incubators
Source: Author's own calculations.

continuum from non-profit to for-profit. The horizontal scale represents incubator objectives ranging from traditional (new venture creation, job creation, economic development) to specific (transfer of technology, sector-specific objectives). The grey line represents the total number of incubators established through time as depicted in Figure 10.2. Leblebici and Shah (2004) identify the period 1960–84 as the origin of the industry.

First generation incubators aimed at job and venture creation that are catalysts for economic development (Hackett and Dilts 2004b; Leblebici and Shah 2004). The first phase of growth (10 per cent yearly growth rate on average) of the incubators in the US between 1984 and 1990 was a reflection of the economic downturn of the US economy. After a slow diffusion of incubators from the 1960s to 1984, massive public funding was directed to incubators during the first phase of growth, as a remedy for unemployment and firm failure (Leblebici and Shah 2004). The movement towards creating university technology incubators and the favourable image of incubators enhanced this trend. The number of incubators rose to 400 in the US only and industry experts estimated that there were about 1,000 incubators around the world by 1990 (Allen and McCluskey 1990). These factors played an important role in sustaining incubators as an important tool for promoting entrepreneurship. In this period most incubators were publicly financed and aimed at economic development by industrial renewal and creating jobs.

The incubation industry reached a level of maturity in the following five-year period. In the mid-1990s the number of newly established incubators in developed countries slowed down. This was mainly because of the problems in the incubation process. Most incubators were providing similar low quality business support services without any exit restrictions. The credibility of incubators was restored in the second half of the 1990s which could be labelled as deepening of the industry. The development of the new high-tech economy stimulated this process and led to the creation of sector-specific incubators (Hackett and Dilts 2004b) and virtual internet incubators that aimed at stimulating networking among high-tech start-ups (e.g. Nowak and Grantham 2000).[3] Another factor that enhanced this second growth wave was the adoption of the concept in developing countries such as China, Brazil, India, Malaysia, and Turkey as can be seen in Figure 10.2. One-third of the existing incubators in developing countries were established from the early 1990s to 2000. It is now estimated that around 40 per cent of all existing incubators are in developing countries (Centre for Strategy and Evaluation Services 2002).

10.2.1.2 FINANCING INCUBATORS

The next generation incubators are expected to be for-profit and sector-specific. The founder and the first director of the NBIA, Carlos Morales, argues that for-profit incubators are expected to grow to about half the total number of all incubators in the coming years (paraphrased in Nowak and Grantham 2000). Becker and Gassmann (2006) show the increasing trend of corporate incubators with a thorough typology of for-profit incubators. For-profit incubators could be organized as independent entities or they could operate under a parent corporation. Their major purpose is to generate revenue for their owners which could be done in various ways such as charging fees for services provided or investing in the businesses of individual tenants. For-profit incubators also support corporations in achieving economic goals such as developing innovative products and processes.

Given the market failure argument behind establishing incubators the label 'for-profit' might sound odd. However, the idea of for-profit incubators is compatible with the market failure argument. New and small ventures have problems to reach resources, funding, and knowledge. Firms may choose to receive assistance for specialized services provided by an incubator, but in return have to give up shares or pay for these services. In this case all the services are provided and organized within an incubator and the only difference is that services have monetary costs. The benchmarking study in EU countries shows that for-profit incubators constitute about 25 per cent of all incubators in Europe, ranging from 38 per cent in Italy to 18 per cent in France (Centre for Strategy and Evaluation Services 2002). In the US this number is about 15–20 per cent.

In developing countries most incubators are still funded by the government and the for-profit idea is yet to develop. There is no clear information on how much governments spend on incubators because most funding available for incubator promotion is integrated into other major funding programmes.[4] For instance, many incubators are supported by the EU through EU structural funds and regional development funds. International institutions such as the World Bank and UNDP have only a modest role in financing incubators in the developing world.[5] However, a recent World Bank initiative led to the establishment of *info*Dev, which specifically aims at stimulating innovation in developing countries by establishing incubators that assist new technology-oriented ventures especially in information and communication technology (ICT) (Scaramuzzi 2002).[6] Section 10.4 gives more information regarding how incubators are financed in developing countries.

10.2.1.3 INCUBATION IS AN EVOLUTIONARY IDEA

As stressed earlier this chapter suggests that the definition and the forms of incubators have evolved through time according to the needs of the firms and the economic climate. Figure 10.3 reflects this with a rough correlation between the introduction of new incubator models and the number of incubators established each year. In this sense we contradict Grimaldi and Grandi

(2005) who identified two incubating models, comparing four different models of incubators: business innovation centres, university business incubators, independent private incubators, and corporate private incubators. Model 1 aims at reducing start-up costs for new ventures anchored in the old economy, targeting local markets by providing various business support services. Model 2, on the other hand, aims at accelerating the start-up process of high value added high-tech companies by providing both technological and management assistance. They argue that the emergence of the second model does not represent a linear evolving process or an improvement of Model 1. Our view is more in line with Leblebici and Shah (2004) who argue in narrative theory perspective that the concept of non-profit incubators in the 1970s has been reinvented in the late 1990s for the new techno era. The long-term evolution of incubators reveals that the concept has evolved from a simple tool for economic development into a high-tech, sector-specific, and increasingly profit-oriented tool to promote entrepreneurship.

10.2.2 *Value added contributions of incubators*

There are a number of studies that evaluate the usefulness of incubators by assessing their value added contributions. Appendix table 10.1 lists research that assesses incubator performance for various countries. One fact that makes the assessment challenging is the selection of appropriate criteria. On what grounds can an incubator be labelled as successful? Answering this question requires a brief summary of the aims of establishing incubators. Incubators are established for different reasons:

1. *To reduce start-up and early stage operational costs, and the risk of doing business by providing a protective environment for start-ups.* (Similor and Gill 1986; Allen and McCluskey 1990; Lalkaka 2002). Most incubators offer managerial and administrative assistance as well as physical infrastructure to their tenants. Previous studies showed that incubator services are important for tenant firms (Mian 1997). For instance, Adegbite (2001) argues that one of the main reasons behind the low performance of Nigerian incubators is poor and insufficient incubator services. Especially managerial assistance could be an asset to entrepreneurs who lack managerial skills.
2. *As a means of regional (technology) development policy* Incubators were used as an effective policy tool in various countries for reducing unemployment, new job and venture creation. See, for instance, Thierstein and Wilhelm (2001) for the German case, Roper (1999) and Frenkel, Shefer, and Miller (2008) for the Israeli case, and Abetti (2004) for the case of Finland.

3. *Enhancing university–industry collaboration* via *university incubators*. Especially in the mid-1990s incubators were established with the aim of increasing commercialization of research and transfer of technology. See, for instance, Mian (1996a), Phillips (2002), and Rothaermel and Thursby (2005a, 2005b) for the US case, and McAdam and McAdam (2008) for UK and Ireland. University incubators also serve as a role model for university students and act as an in-house (part-time) employment opportunity for students (see Akçomak and Taymaz 2007 for the Turkish case).
4. *Stimulating networking among firms* (e.g. Sweeney 1987; Allen and McCluskey 1990; McAdam et al. 2006). Tenant firms can benefit from peer group effects. The idea is based on synergies among entrepreneurs who share similar problems, businesses, and work environment. For instance, Hansen et al. (2000) argue that among the existing incubator models, the networked incubator (incubators in which networking is organized and deliberately fostered) is likely to be more successful. In a similar manner, Hughes, Ireland, and Morgan (2007) argue that firms' success is related to strategic networking not to their mere presence in an incubator. Tenant firms network to access resources and to acquire knowledge.
5. *Reversing or preventing brain drain*. For instance, in Israel high-tech incubators were effectively used as a tool for absorbing immigration (Roper 1999; Frenkel, Shefer, and Miller 2008). Between 1989 and 1995 more than 11,000 high-skilled scientists and engineers emigrated from the former Soviet Union, some of whom were employed in incubator firms. Incubators can also help scientists to commercialize their work and to increase the financial means of scientific research. For instance, one particular goal of the Zelenograd Scientific and Technology Park in Russia is to make scientific work financially worthwhile to attract scientists back. Russian science has faced a within country 'brain drain' in the sense that most Russian scholars gave up scientific research for more profitable non-scientific work such as managing western retail stores in Moscow (Bruton 1998). Similarly, China established 'Innovation Parks for Returned Scholars' to attract talented researchers and students who live abroad. Various subsidies are provided for returned scholars to set up high-technology-oriented businesses in China (Chandra 2007).

Looking at the above list of diversified goals it is not easy to assess incubator success. One issue we have mentioned earlier is the criterion for performance. Is it survival of the firm, generation of employment, growth of sales revenue, or innovativeness? Such indicators have been employed previously in

Table 10.1 Summary findings of the literature assessing incubator performance

	Survival	Sales growth	Employment growth	Innovativeness
Physical infrastructure	+	+ / O	O	O
Management support	+ / O	O	O	O
Administrative support	+ / O	O	O	O
Incubator image	+	+ / O / –	O	O
Financial support	+ / O	+ / O	+ / O	O
Networking with university	O	O	O	+ / O
Networking with business	+ / O	+ / O	+ / O	+ / O
Networking with incubator firms	O / –	O / –	O / –	O / –

Notes: (+) positive effect, (–) negative effect, (O) no particular effect.

assessing the performance of incubator tenants (see Table 10.1). So far there is no consensus among researchers regarding which indicators to focus on. Another related issue refers to the goals of the stakeholders. Different parties with different objectives are involved in an incubator and a success indicator for one stakeholder may not be an indicator of success for another. For instance, in a university incubator, the survival rate is an important concern for the incubator management. Therefore, a high survival rate can be considered as a success criterion. However, if this is not accompanied by an increase in the employment of graduates of the university in which the incubator is located, it may be considered as a failure from the university's point of view. Moreover, since the opportunity cost of incubation is unknown, it is very hard to determine whether an incubator is successful or not. Given the above goals we summarize the findings of the literature assessing the performance of incubators in Table 10.1.

The findings of this literature for various countries are presented in Appendix table 10.2. Here, we highlight several points that have emerged from these studies. First, most evaluation studies are conducted in developed countries. There are only a few studies that quantitatively evaluate incubator experience in developing countries (e.g. Hsu et al. 2003; Akçomak and Taymaz 2007). Second, qualitative (case) studies tend to be more optimistic regarding the usefulness of the incubators. Most studies argue that lowering start-up costs and providing assistance increase the likelihood of survival and create ventures and jobs. However, this is only one part of the story as no such information is collected about what would have happened if the entrepreneurs had decided not to locate the firm in an incubator. Statistics show that in general survival rates are much lower for off-incubator firms. However, some studies report evidence that does not support overall statistics. For instance, in Germany it was found that only 3 per cent of firm owners would not have started their business without the existence of incubators (Tamasy 2007). Similarly, in Turkey among a sample of 48 incubator firms only two stated that the company would have not existed if it were not located in an incubator (Akçomak and Taymaz 2007). On the other hand, quantitative studies tend to report mixed evidence on performance indicators such as firm sales and employment growth and innovativeness. Researchers generally agree that incubators do have an impact on employment and sales growth, but not on stimulating innovativeness in terms of patents, product and process innovation. In a study to assess technology transfer and the commercialization of research, Phillips (2002) found that technology business incubators are only marginally beneficial and the level of technology transfer is much lower than expected. Third, there are only a couple of studies that pursued longitudinal analysis (Rothaermel and Thursby 2005a, 2005b). To assess the impact of incubators thoroughly, information on after-incubation is crucial. Unfortunately data availability significantly hinders this type of analysis. Finally, almost all studies assessing the performance of incubators are based on assessing firms rather than the incubator itself. However, the differences between firms within an incubator and between on- and off-incubator could reflect the motivations of the firms as well as the benefits of an incubator (Löfsten and Lindelöf 2002). For instance, it may be the case that incubated firms perform better regarding economic or technological indicators because entrepreneurs of these firms might be more qualified (Pena 2004).

10.3 Country cases

This section provides experiences of four developing countries in business incubation. We have chosen China and Brazil because these two countries together host about 1,000 incubators and are ranked third and fourth respectively (in terms of number of incubators) following the US and Germany. To match China and Brazil we have selected India and Turkey. China and India are two emerging developing countries that are comparable in terms of their size, GDP growth, and potential they provide for future development. On the other hand, Brazil and Turkey are accepted to be two promising emerging market economies and display certain similarities such as institutional landscape towards democracy, fully functioning market economy, macroeconomic instability, etc. All four countries (i) have to promote an entrepreneurial risk-taking culture, (ii) have inhibiting institutional obstacles to entrepreneurship, (iii) have underdeveloped venture capital (VC) markets, and (iv) face similar difficulties in managing incubators. Given these factors, the evolution of incubators, their goals, incubation models, and financing structure display considerable differences among countries. Appendix table 10.2 summarizes the incubation experiences in four countries in comparison to the US.

10.3.1 *Brazil*

Brazil has more than 30 years of experience in incubation. The Brazilian incubator movement took off in the 1980s with the collapse of the military regime and the emergence of democratic institutions. The first incubator was established in 1986 and within ten years this number reached 40. The growth of the incubation business was rather slow in the first decade mainly due to inconsistencies between the national programme and the commitments to grow. According to an early assessment by Lalkaka and Bishop (1996) most incubators were located in a university and more than 80 per cent of the tenants were spin-offs from academia and other companies. Universities played a vital role in establishing incubators and they promoted the idea of incubation until incubators were accepted as a tool to promote entrepreneurship. In this sense one can argue that incubators in Brazil flourish as a product of a bottom-up process (Etzkowitz, Carvalho de Mello, and Almeida 2005; Chandra 2007). At this initial stage the main weaknesses of the incubators were poor relations with academia and insufficient business support services. There were cases where incubators did not provide business support and consultancy services at all. Moreover some incubators provided very poor physical and operational infrastructure. These problems were exacerbated by institutional constraints to entrepreneurship, such as bureaucracy and insufficient risk-capital funding.

Because of such deficiencies the idea of incubators developed slowly as a complement to other instruments, such as Micro and Small Business Support Services (SEBRAE) that has various support mechanisms for small firms, Núcleos de Inovação Tecnológica (NIT) that aims to assist researchers in the transfer of technology and intellectual property, and the Ministry of Science of Technology that coordinated activities to provide small-scale financial support and venture funds for small and new technology-oriented ventures. The need for a national business incubation strategy made the Industry Technological Training Support Programme (PACTI) cooperate with the National Advanced Technology Enterprise Promoter Entity (ANPROTEC) in establishing a sustainable strategy to support business incubation.

The incubation industry lacked a national strategy for quite a long time. This deficiency, contrary to expectations, established a flexible environment that produced different incubator models for different aims and helped Brazil to create a less costly development model (Etzkowitz, Carvalho de Mello, and Almeida 2005). Lack of national policy enabled an environment of experimentation. It is because of this experimentation that incubation in Brazil developed in an innovative way. Incubation models, approaches to incubation, and types of incubators differed to a considerable extent and this diversity in turn fed further developments in incubation. In the mid-1990s these efforts supporting the concept of incubation paid off quickly. Between 1995 and 1997, 33 new incubators had been established which outnumbered the cumulative number of incubators that were established since 1986. There are now about 400 incubators operating in Brazil more than half of which are technology business incubators. According to Almeida (2005) in 2003 there were over 1,000 tenant firms with more than 15,000 employees.[7] The latest figures from ANPROTEC show that the mortality rate of incubated firms is 7 per cent compared to about 50 per cent within new ventures that are not incubated (Oliveira and Menck 2008).

Incubators in Brazil are generally linked to universities and financed by various governmental and non-governmental sources, such as the National Incubation Support Program (PNI) that supports the creation of new incubators and assists the existing ones to expand—private companies and the Federation of Industries. The Federation of Industries in São Paulo is actively involved in operating about ten incubators (Chandra 2007). The active contribution of various actors (for-profit, non-profit, and universities) is a distinct feature of the Brazilian incubator scene. Therefore incubators in Brazil are reflections of synergies (a triple helix) among the university, industry, and the government (Etzkowitz, Carvalho de Mello, and Almeida 2005). This synergy even had indirect repercussions on innovation in general. For instance, the Innovation Law in 2005, that significantly improved transfer of technology and commercialization of research, is an indirect product of this synergy.

Another interesting feature of incubators in Brazil is the innovative approach to incubation models. Different incubation models emerged because of local needs and attempts to alleviate poverty. Many incubators were established as a remedy to unemployment, aiming solely at job creation especially in traditional sectors such as agricultural equipment, furniture, and textiles. There are even incubators that are specialized to foster entrepreneurship in cultural activities such as music, art, and cinema industry (Scaramuzzi 2002). Recently, PNI has initiated a very interesting programme that aims at sharing information, experience, and expertise among incubators. According to this programme older and reputable incubators are asked to help smaller incubators (Chandra 2007). In other words, older incubators will incubate younger incubators.

10.3.2 *China*

China is a relative newcomer to the incubation business. The first incubators were established in the late 1980s but the growth of the industry has been tremendous. In the first ten years 100 incubators were established incubating about 1,400 firms. The recent estimates show that there are now more than

500 incubators in China sheltering over 600,000 employees (Chandra 2007). China has achieved an outstanding success regarding the expansion of the incubation programme. As in most other cases the government played a predominant role. Incubators in China offer services such as low-cost office space, business support services, and networking opportunities. However, they are much bigger in size compared to their counterparts elsewhere (Scaramuzzi 2002). An average incubator shelters 60 to 70 firms and it is not surprising to see incubators with more than 150 new ventures (Harwit 2002).

The entrepreneurs in China, like in any other developing country, lack financial resources and managerial skills. However, in the Chinese case there are two additional barriers to entrepreneurship: (i) the financial and institutional system which is solely dominated by the government, (ii) the risk-averse cultural values. The risk-averse culture not only inhibits new venture formation but also affects seed fund decisions of the banking and financial sector. The Chinese incubators have to overcome typical barriers to entrepreneurship and moreover have a mission to transform societal values towards a risk-taking entrepreneurial culture. In this sense the Chinese government used incubators as a policy instrument to create markets especially for high-technology products. Incubators in China are financially supported by the government via the Torch High Technology Industry Development Center, under the Ministry of Science and Technology (MOST). The Torch Center is involved not only in constructing incubators with modern infrastructure and equipment but also in operating these incubators. The government provides generous seed-capital funds for start-ups and funds for R&D and innovation for small and new technology-oriented firms. But the main reason behind locating firms in incubators is the low-cost services such as low rent, which can be 50 per cent below market rent (Harwit 2002).

What makes the Chinese case special is the emphasis on innovation and technology. The incubation industry is strategically designed to play a role in China's transition to a high-technology-driven market economy (Harwit 2002). As an illustration, incubator policy was a core ingredient in the tenth five-year plan (2001–5) of MOST. US$6 million were provided only for constructing incubators (Chandra 2007). As of 2002 about 40 per cent of the high-tech firms in development areas were housed in incubators (Harwit 2002). The Chinese experience lacks sound and quantitative assessment but the estimates show that the transformation rate of scientific achievements into production increased on average from 30 to 70 per cent. Moreover the survival of high-tech ventures increased from 30 to 80 per cent because of incubation. Currently there are various types of incubators with different models some of which are innovative in set-up, such as the incubators for returned overseas scholars that aim to provide finance and assistance for scholars who live abroad but who plan to implement their creative ideas in China (Ling et al. 2007).

A particular weakness of the Chinese incubation programme is the dependence on the government. Incubators are financed and managed to a greater extent by the government and this dependency hinders self-sufficiency of incubators in China. Incubators are required to reach self-sufficiency in three years but they seldom accomplish this target. This dependency even affects the market orientation and behaviour of tenant firms. Because the exit criterion is not implemented strictly most firms tend to linger around and continue their risk-averse structure (Harwit 2002; Chandra 2007). Moreover, Chinese incubators are found to be less active in providing a variety of internal and external financial services (Chandra, He, and Fealey 2007). Another weakness of the programme is the focus on the 'hardware' aspect (Lalkaka 2002). The government has invested a massive US$1.6 billion to construct incubators. However, the intangible services that are provided by the incubators are poor in quality and management skills of the managers—most of whom are government officials and have no market experience—are low. The size of the incubators is another factor that reduces the quality of the services. Harwit (2002) reports cases where this poses serious problems as it is not easy to provide similar concern for all tenants if the incubator has about 100 tenants.

10.3.3 *India*

India has about eight scientists and technical personnel per 1,000 inhabitants. This number is low compared to countries such as Russia (113) and South Korea (22). However, absolute numbers are more revealing to understand the potential of India: there are about 10 million technical personnel, a pool which is almost ten times bigger than most countries. Starting from the 1950s, the Indian government has initiated some policies to leverage this talent, such as establishing prominent universities and research institutes, providing tax exemptions to new ventures, improving financial and VC markets, and the establishment of the National Science and Technology Entrepreneurship Development Board (NSTEDB) in 1982. However, it is still not possible to argue that India is benefiting optimally from this large talent pool. It is true that India has built a massive science, technology, and research network composed of hundreds of universities, laboratories, and institutes, but the reflection of this in entrepreneurship has been surprisingly limited. Most entrepreneurs lack necessary capabilities to manage a business, have problems in networking, and last but not least, lack financial resources. Despite improvements, the financial institutions provide capital on stringent conditions, such as high interest rates and a demand of 110 per cent collateral security. This leads to a financing gap which discourages entrepreneurs. Incubators could therefore be a stimulus in turning talents and new ideas into new ventures.

The incubator movement in India took off in the late 1980s as a complementary policy tool aiming at promoting entrepreneurship and stimulating new venture creation. The take-off was slow because the first incubators were financed by the United Nations, but lacked government support (Lalkaka 2002). This trend has been continuing since then and it seems that the incubator movement is not picking up in India. Currently there are about 50 incubators in India (15 of which are technology business incubators) compared to about 400 incubators in China and about 300 in South Korea. Higher education institutes are hesitant to support incubators and it is not easy to raise public awareness of the role that incubators could play. There are recent initiatives of the Department of Science and Technology together with other programmes like Entrepreneurship Development Cell and Science and Technology Entrepreneurs Park (STEP) to set up more incubators (Saravanan 2007). These recent initiatives proposed by the government are (i) incubation funds that provide seed money to entrepreneurs, (ii) tax exemptions for services provided by incubators, (iii) priorities for incubated enterprises in financial markets, and (iv) supporting the establishment of incubators in the fields of chemicals and biotechnology (implemented by the Department of Biotechnology). It is best to complement these efforts with campaigns that promote the concept of incubators. For instance, the global forum on business incubation commenced in Hyderabad in 2006, hosting 300 practitioners from 60 different countries which was a successful promotion attempt to raise awareness.

Beside the government of India and the UN, the World Bank is also active in the incubation industry. *Info*Dev, a partnership of international development agencies, has been financially supporting five incubators countrywide: (i) IndiaCo (Mumbai), (ii) TREC-STEP (Tiruchirappalli), (iii) SRISTI (Ahmadabad), (iv) TeNeT (Chennai), and (v) VIT-TBI (Tellore). The amount of the grant reaches up to US$1.1 million. Incubators under this scheme function like any other technology business incubator. They offer office space, ICT facilities, and other incubator services to their tenants. Moreover, it is possible in some cases to receive financial assistance in the form of seed funding directly from the incubator (*Info*Dev 2008).

What makes India rather an interesting case is that incubators constitute only a small part in policy that promotes entrepreneurship. By means of various different programmes and mechanisms organized by the NSTEDB more than 100,000 researchers and students have received training in entrepreneurship since the beginning of the 1990s (Saravanan 2007). By any standards this is a big number and an achievement to be applauded.

There is not a study, to our knowledge, that assesses the achievements of incubators. The estimates and limited evaluation studies show that the incubated firms grow much faster than their non-incubated counterparts and the survival rates are about 80 per cent, compared to only about 40 per cent for new ventures that are not incubated.[8] An effort should be made to conduct sound assessment studies, which could be a valuable asset for practitioners and policy makers worldwide.

10.3.4 *Turkey*

As one can see in the chapter by Bascavusoglu-Moreau, SMEs (99 per cent of all firms) have an important place in the Turkish economy. The government has employed various policy tools to assist SMEs such as direct financial support, R&D subsidies, and tax allowances. As a policy tool in this direction, the incubator movement in Turkey started in the 1990s with a particular interest in technology business incubators (TBIs). The concept of an incubator was appealing, because the risks of establishing and maintaining a high-technology business were manifold in an environment characterized by macroeconomic instability such as in Turkey.

Incubators in Turkey are established by the Turkish Small and Medium Sized Industry Development Organization (KOSGEB), which is a non-profit, semi-autonomous organization (under the Ministry of Industry and Trade) with the objective of improving the conditions of SMEs and enhancing their competitive capacity. Within KOSGEB, there are three different schemes that can be identified as an incubator: Enterprise Development Centres that function as traditional incubators, Incubators Without Walls, and Technology Development Centres (TEKMERs) that function as university incubators. TEKMERs aim: (i) to help people who are trained in scientific and technological fields to become entrepreneurs; (ii) to foster the creation of new technology-based enterprises; (iii) to foster the commercialization of R&D; (iv) to strengthen university–industry cooperation; and (v) to assist other policy initiatives that focus on the diversification of regional economic activities.

TEKMERs select entrepreneurs/firms after an assessment of their projects on the basis of technological orientation and innovativeness. Therefore the majority of the incubator tenants are limited liability firms that are specialized in computer, software, electronics, medical appliances, and automotives. As of 2008, a total of 650 firms have benefited from incubator services, most of which are new technology-oriented ventures. Currently there are about 40 different types of incubators, 15 of which are technology incubators. Incubators provide office space and necessary supplies and equipment to tenant firms. Moreover, firms can benefit from diversified financial assistance initiatives. These include: (i) a support of (in the form of loan) up to about €120,000 for supplies, equipment, and prototype development; (ii) initiatives ranging from professional consultancy support to rent support, without repayment up to a total of about €30,000. Firms can also take advantage of the facilities provided by the universities where the incubator is located. The chief advantage of incubators is that firms receive a considerable amount in tax reductions.

In 2003, the performance of incubator firms was evaluated by means of conducting a survey in six incubators in Ankara and Istanbul (Akçomak and Taymaz 2007). The study was based on a matched sample technique, comparing incubator firms with firms located elsewhere in terms of performance factors.[9] The main findings of the study are summarized below.

The incubator firms perform better than their counterparts located elsewhere regarding employment growth and sales revenue growth. The majority of on-incubator firms stated an increase or no change in employment. Only about 15 per cent indicated a decrease in total employment. Moreover only about 5 per cent of the firms stated a decrease in the employment of R&D personnel. On the contrary, more than one-third of firms that are located elsewhere faced a decrease in employment (total employment or R&D employment). A considerable part of the increase in employment is due to employment from the local market. For instance, 75 per cent of the firms stated that there is at least one employee who graduated from the university where the firm was located. Similarly, 29 out of 48 firms have at least one entrepreneur (founder of the firm) who graduated from the university in which the firm is located. This favourable situation also holds for growth in sales revenue. More than 60 per cent of the incubator firms faced an increase in their sales revenue. The study found no statistically significant difference between on- and off-incubator firms regarding innovation (owning trademarks, patents, product, and process innovation, etc.).

What makes incubator firms more successful in growth of employment and sales revenue? Three factors may explain this difference. First, it might be the case that on-incubator entrepreneurs are more qualified which makes the firms they establish more successful. Second, financial opportunities and incubator services may determine the success. Third, incubator firms are more likely to interact with other firms and universities and therefore can benefit more from network externalities. The findings reveal that entrepreneur characteristics and networking do not explain the differences between on- and off-incubator firms. Incubator firms are linked to universities, but most of the interaction is in the form of informal contacts and could easily be explained by proximity to university location. Concerning the second argument, financial support is found to be critical for incubator firms. The detailed interviews revealed that complexity of the application procedure and deficiencies in the implementation hinder the full impact of the financial support mechanisms. An overall assessment of all incubator services shows that 60 per cent of the firms found incubator services critically important for the survival. However, the main reasons of locating the firm in an incubator are the university's image and tax benefits. Incubator services and networking opportunities are surprisingly among the least important reasons.

The absence of VC initiatives, low levels of business networking, lack of marketing initiatives, and inadequate business support mechanisms appear to be among the major problems of the incubators in Turkey. The main source of funding in the start-up stage is own capital accumulation of the entrepreneur. VC and spin-off processes are vital for the long-term success of incubators but there is little improvement in VC markets in the last decade. The low level of business networking and interaction is one of the main weaknesses of incubators in Turkey. Only 10 per cent of the on-incubator firms stated that the reason for locating the business in an incubator is to network with similar firms. Unfortunately, enhancing business networking in incubators in Turkey is not an easy task because firms do not consider networking as an essential strategy. Entrepreneurs deliberately chose not to network with other firms due to reasons of secrecy, protection of sensitive information, and critical personnel. Last but not least, one of the main deficiencies of the entrepreneurs in incubators is their lack of experience in management. Most firms in incubators do not have viable business plans and the management support from the incubator is deemed insufficient.

10.4 Requirements for successful incubation

The requirements for a successful incubation process could be analysed in a simple demand–supply framework. The entrepreneurs and new ventures constitute the demand side. The entrepreneurial actors lack managerial and marketing (and to a certain extent technological) know-how and demand (specialized) tangible and intangible services, administrative support, easier access to resources, and access to business networks. The agents on the supply side are the stakeholders in incubators. Business communities, local governments, and universities support new ventures and provide business services and technological assistance through incubators for different purposes. Success depends on elements of supply (exogenously), i.e. the management and funding of incubators, the quality of services provided, clear-cut well-defined strategies and purposes, and elements of demand (exogenous), i.e. how receptive are the entrepreneurs for capability building. Incubator success depends on how well the sponsors and incubator management can meet the specialized demand from the entrepreneurs.

In the previous sections we have reviewed the literature evaluating incubators and presented four country examples from the developing world. What can we learn from these experiences? How should an incubator policy be organized to increase effectiveness? In this section we summarize the requirements for successful incubation by giving brief examples from experiences from the developed and developing world.[10] Eight conditions for successful incubation are identified.

- *The purpose and the mission of the incubator should be clear*: in order to provide useful assistance incubators should know their mission clearly (Lalkaka and Bishop 1996; Kim and Ames 2006). Incubators are established for different purposes from transfer of technology to job creation. Knowing the mission clearly is important for the incubator-manager in providing the right service-mix for entrepreneurs. It is also vital for the selection of firms, entry-exit decisions, and their implementation. Countries should not set up incubators without a strategic plan just for the sake of establishing incubators.
- *Incubators should set clear selection, entry, and exit criteria*: the existence of clear criteria for entry is positively associated with the success of the incubator (e.g. Hackett and Dilts 2004b; Totterman and Sten 2005). There are many cases in developing countries that report problems in the implementation of this very simple rule. For instance, studies for Nigeria (Adegbite 2001), Malaysia (Yunos 2002), China (Chandra 2007), and Turkey (Akçomak and Taymaz 2007) have stated that most incubators fail to meet this criterion. In China and Nigeria this problem is exacerbated because tenant firms tend to remain in the incubator even though the incubation period has expired. This is also true for Turkey. There are even arguments that the incubation period should be flexible to maximize firm benefit because a longer incubation period might have negative impact on some tenant firms (e.g. Hytti and Maki 2007). Incubation provides a secure environment and may obstruct firms in developing certain skills, such as competition and marketing which are generally acquired through learning by doing.
- *Incubator-managers should be qualified, preferably with business experience*: managerial capacity and skills of incubator-managers are critical for the success of the tenant firms. In almost all developing countries this is a major problem. The incubator-managers are not qualified (Kim and Ames 2006), do not have a business background (Lalkaka 2002), and are not business-oriented (Tamasy 2007). Therefore, incubators are not able to provide the right service-mix for their tenants. They are also not able to embed firms into networks and have difficulties in evaluating and developing the business plan of their tenants. There is a supply shortage

of qualified managers especially in countries where the incubator movement expanded rapidly such as China and South Korea. The popularity of incubators increases the demand for managers but it is not possible to train and generate qualified incubator staff at the same speed.

- *Monitoring firms is essential for success*: monitoring and screening tenants to assess their development is essential for incubation (Hackett and Dilts 2004b; Aerts, Matthyssens, and Vandenbempt 2007). Tenant firms could be assessed on their performance regarding employment, sales growth, financial position, or innovativeness. Regular monitoring reveals the state of firms, their specific needs, and their success potential. It is the manager's job to be pro-active in monitoring. Firms should be evaluated carefully on the basis of their management skills and financial strengths before admittance to an incubator. This is an easy solution to minimize firm failure (Aerts, Matthyssens, and Vandenbempt 2007).
- *Services that are provided by the incubator should be strategically selected*: incubator services help entrepreneurs to avoid start-up cost and in some cases avoid start-up bureaucracy (e.g. the case of Brazil). There are direct and indirect externalities from pooling resources that increase efficiency. For instance, facilities are shared by tenants so money can be invested in other useful activities (e.g. Hackett and Dilts 2004b; Chan and Lau 2005). Incubators should strategically select the services provided and monitor their use. Self-evaluation of services provided is not a sufficient criterion to determine the right service-mix to be offered for tenant-companies. Services that are not provided can be also essential for firms (Allen and Rahman 1985), thus monitoring should also address services that are not provided to see whether there is need for such services. Self-assessment is important to reveal where the incubator stands.
- *Intangible services are much more important than tangible services*: most incubators in developed countries provide intangible and specialized services for tenant firms (e.g. networking strategy, assistance in making business plans, marketing, etc.). In developing countries the emphasis is on tangible services, such as office space, infrastructure, and laboratories. This does not mean that these services are not important. On the contrary, in developing countries tangible services are of utmost importance. However, in many cases (e.g. China, Malaysia, Nigeria, and Turkey) there has not been an optimal mix between the tangible and intangible services that are offered by the incubators. Given the needs of the contemporary world, incubators should be able to provide high quality and specialized intangible services.

- *Networking adds value only if it is a deliberate strategy*: sheltering entrepreneurs under one roof does not guarantee beneficial effects from network externalities. Networking should be a deliberate strategy of the incubator. Only in such a case can one talk of synergies (e.g. Hansen et al. 2000; Bollingtoft and Ulhoi 2005; Chan and Lau 2005). However, some studies report drawbacks of excessive networking and overflow of information. For instance, in a case study in Ireland, McAdam and Marlow (2007) found that being in close proximity to each other might have adverse affects on secrecy (e.g. copying and stealing ideas). In a similar manner, Akçomak and Taymaz (2007) report that tenant firms were hesitant to share sensitive information with other firms in the incubator. Most of these firms are microfirms with less than five employees and firms fear that their projects could be copied by the other firms by transfer of critical personnel and stealing ideas, for example. In these circumstances, trust is a critical factor that enables information exchange.
- *The incubator should be able to self-sustain its operations*: in most developing countries incubators are dependent on government funding. Tenant firms are expected to be self-sustaining after spending three or four years in an incubator. Given this one could also expect incubators to become independent. Self-sustenance requires networking, strategic planning, and diversified funding sources and forces incubators to become more business-oriented. Just as incubators force entrepreneurs to become more competitive, the government may facilitate (or induce if necessary) incubators to become self-sustainable.

10.5 Concluding remarks

The country experiences discussed above reveal that contemporary successful incubators provide a wide range of services, focus more on intangible business services, and employ qualified managers and support staff. To match the needs of the current market economy and the requirements of the firms the concept of incubators has been reinvented continuously and has evolved through time. The main weaknesses of incubators in developing countries are: (i) focus on tangible services; (ii) reliance on government both in terms of promotion and funding; (iii) lack of qualified personnel and skilled managers; and (iv) lack of planning and creativeness in solving problems and providing services.

The first three of the above weaknesses are observed in many developing countries but are not easy to tackle. These problems are, so to say, endogenous to the system. For instance, the provision of high quality tangible services requires highly skilled managers and personnel which in turn depend on the supply of incubators and the general policy on education and training. The government can chose to invest lavishly in creating incubators but this investment has an opportunity cost which could well be a forgone investment on skilled employees in innovation management. Hence the incubation industry would end up with less skilled personnel who are not productive and knowledgeable in providing intangible resources. For example, the case of South Korea fits well into this framework. The extreme increase in the number of incubators created a shortage of experienced managers and had a bad impact on the quantity and quality of the services provided by the incubators. The Chinese experience is another good example of policy inconsistency. Salaries of around €300 per month at Fudan University make self-employment tempting for university researchers. In some cases researchers spent more than three quarters of their time in incubators to develop their businesses at the expense of teaching and research activities (Harwit 2002). This might seem acceptable if it helps to reach the target of increasing commercialization of research. However, it has formidable drawbacks for the quality of education. Hence there is a need for an optimal policy regarding incubators.

The bottom line is that the incubator policy should be integrated in a framework of policies that promote entrepreneurship in particular, and innovation and economic development in general. Incubators reflect the institutional set-up, creativity, and policy innovativeness in a society. Policy on incubators is neither a quick fix (e.g. Allen and McCluskey 1990) nor a sole cure for all problems regarding entrepreneurship (e.g. Autio and Klofsten 1998). Therefore incubation is not an appropriate tool for policy makers who seek fast short-run solutions.

The mere existence of incubators cannot guarantee people becoming entrepreneurs and cannot induce networking among firms. Therefore it is best to complement incubator policy with policies that encourage entrepreneurship. Take the issue of education for instance. In most developing countries, especially in China, a risk-averse culture hinders entrepreneurship. In such an environment investing in education policy to create a more receptive environment that would promote creativity and entrepreneurship might be a better option than investing blindly in incubators. In a similar manner, in most developing countries financial institutions are also risk-averse and VC markets are underdeveloped. Therefore entrepreneurs heavily rely on personal funds. Moreover, there are institutional and bureaucratic barriers that deter entrepreneurship. In such an environment improvements in financial institutions are the first-best solution to start promoting entrepreneurship.

Given the evolution of the concept of incubators the policy towards incubators needs to be flexible and innovative itself. There is no problem in implementing incubator models that are borrowed from developed countries as long as they are modified according to the economic climate, values, and the institutional landscape of the receiving country. Most successful incubators display a creative and innovative character in approaching problems of tenant-companies. Incubators encourage firms to become innovative and competitive. Such a mission can be pursued only if incubators themselves become competitive, business-oriented, and innovative. There are cases where innovative models of incubators emerge as a response to local needs. The social incubator and the incubators specialized in art and culture in Brazil, and the returned scholar incubators in China are good examples of novel incubator models.

Assessing incubator performance is not a simple task and the literature presents conflicting findings. Sound evaluation necessitates good data but data gathering is costly. However, evaluation should be a routine task of incubator management rather than being a once and for all luxury. It is the only way to assess the quality of the services provided to tenant firms. Another reason that makes evaluation hard is the opportunity cost of incubation. As in the Chinese case, creating incubators needs enormous investments which could be invested in another policy tool that also promotes entrepreneurship. For instance, in India incubators are not very popular but NSTEDB organizes different programmes to train thousands of people to become entrepreneurs.

An incubator is an institution as well as a tool to promote entrepreneurship and creating institutions is a first-best but costly solution (Bowles and Gintis 2002; Rodrik 2008). Setting up incubators is a viable but not the sole tool to promote entrepreneurship. This could be achieved in various other ways. Developing countries should seek novel and innovative policy tools as well as modify and adapt the already existing ones.

APPENDIX

Appendix table 10.1 Summary of findings of the literature assessing incubators

Country	Study (year)	Sample and methodology	Indicator for success	Findings
USA	Allen and Rahman (1985)	12 incubators, 56 firms. Response rate (RR) 44%. Descriptive analysis.		Physical services (work space, equipment, etc.), advertising and marketing services, risk management, insurance, and government grants are rated as most useful for survival. 87% of the entrepreneurs stated that they would have started their business without the incubator.
USA	Fry (1987)	76 firms. RR 51%. Descriptive analysis. Comparison with control group.		Incubator firms are more active in business planning, mission statement, strategic plan, etc. than non-incubator firms. Most incubators provide such services but this is more in the form of encouragement and only half of the incubator managers actively participate in the planning of the tenants.
USA	Allen and McCluskey (1990)	127 incubators. RR 70%. Regression analysis (dependent varb.: log jobs created and firms graduated).	Percentage of occupancy, jobs created and firms graduated.	Occupancy rates around 50% show that incubators are not strong real estate ventures. Old incubators with accumulated expertise are more successful than the others. Incubators that admit light manufacturing firms are more successful in job creation. None of the business support services have significant impact on jobs created and firms graduated.
USA	Mian (1996a)	Case study of six university incubators and 47 firms. RR 32%.	12 business and 10 university related services assessed for their usefulness.	Shared office services are more useful than business services, such as assistance in applying for grants, marketing, accounting, etc. Incubator firms benefit from university image, student employees, and university labs and infrastructure. Incubator services have added value contributions.
USA	Mian (1996b)	Case study of two university incubators in University of North Carolina and Case Western Reserve.	12 different dimensions to assess success in creating new enterprises.	Sales of firms increase by about 10 and employment by 4 times within four years. Physical infrastructure, student employees, and faculty consultants are the most important services provided by the incubator (university). University incubator services have positive impact on growth and survival of tenant firms.

USA	Mian (1997)	Case study of four university incubators and 29 firms. RR 35%.	Firm survival and growth and contributions to university mission.	Firms in all four incubators indicate sales and employment growth (150% and 35%, respectively). Incubator firms benefit from university image and universities receive public attention due to press coverage. Student employees are found to be the most beneficial source for firms.
Finland (Helsinki)	Autio and Klofsten (1998)	Case study of two incubators to assess their management policies.		Incubators are embedded in local context and their success could only be analysed in the local settings. 'Success stories' cannot be generalized. Practitioners should be careful in adopting policies that are found to be important in other incubators.
Switzerland	Thierstein and Wilhelm (2001)	Case study of 9 and short surveys of 40 incubators. RR 63%.		The regional economic development aim is missing. This can be partially explained by the fact that contrary to most countries incubators in Switzerland are privately owned in most cases.
Nigeria	Adegbite (2001)	Case study of 7 industrial (business) and technology business incubators.		Neither the business nor the technology incubators achieved their primary goals. Weaknesses: lack of objectivity in admission, insufficient support services. There are organizational difficulties as the incubators are operating under the ministry, as well as poor funding.
Israel	Shefer and Frenkel (2002)	Quantitative evaluation of 21 incubators and 109 firms.	Firm survival, obtaining funding for projects after graduation.	86.4% of the projects in the last three years graduated from the programme and 78% of them were able to secure financial support after graduation. The incubator manager's skills and the selection and monitoring of projects are essential for success.
Italy	Colombo and Delmastro (2002)	45 incubator firms (RR 20%) matched to 45 off-incubator firms. Quantitative analysis on matched sample.	Differences in employment growth and innovative activity (number of patents, etc.).	Italian incubators are successful in attracting high-skilled entrepreneurs. There are no significant differences between on- and off-incubator firms regarding their innovative output. The on-incubator firms outperformed off-incubator firms in employment growth, worker education, participation in EU-sponsored projects, and establishing formal cooperative relations.
Taiwan	Hsu, Shyu, Yu, You, and Lo (2003)	Comparison between Industrial Technology Research Institute (ITRI)	Satisfaction with business support services and effect of clustering.	ITRI incubator tenants are more satisfied with incubator services when compared to firms in other incubators. It was found that industrial clustering is important for the

Continued

Appendix table 10.1 continued

Country	Study (year)	Sample and methodology	Indicator for success	Findings
		incubator (16 firms 50% RR) and other incubators (34 firms 16% RR).		development of the incubator which implies that complementarities are important in local economic development.
Finland	Abetti (2004)	Case study of 5 incubators among 16 incubators in Finland. A general assessment for the Helsinki region.	Survival rates and job creation, sales growth.	The survival rates reach to 95%. The incubators receive little funding from the government but are able to create high-skilled cost effective jobs (government subsidy per created job is 6,450 which is much less than the welfare costs per person in Finland). Average sales growth rose by 160% per year during and after incubation.
Hong Kong	Chan and Lau (2005)	Case study of 6 incubator firms.		Rental subsidies and office spaces are important for entrepreneurs. The training programmes are also found to be useful for incubator tenants. On the contrary, firms indicate that they do not gain benefits from clustering.
Spain (Basque country)	Pena (2004)	Quantitative evaluation 114 firms in 9 incubators.	Sales, employment, and profit growth.	Human capital of the entrepreneurs has significant impact on sales and employment growth. Most incubator services have no impact on performance indicators.
USA and South Korea	Lee and Osteryoung (2004)	Questionnaire for firms, incubator managers in South Korea and US (only university incubators). RR 62% and 45%, respectively.	Perceived importance of 14 items measuring critical success factors	There is significant difference between US and South Korean respondents regarding the role of the incubator's strategy (clarity of goals, management, entry/exit policies, business plans, etc.) on the performance of the incubator. Respondents from US give more importance to these factors. Financial support and business networking have more perceived importance than other factors.
USA	Peters, Rice, and Sundararajan (2004)	Secondary data on 43 US incubators. Additional survey for managers.	Number of graduate firms.	Graduation rates are higher in incubators that offer training and that provide an accessible network (consultants, scientists, customers, other business firms, etc.). Non-profit incubators are found to be more successful than other incubator types regarding the number of firms graduated.

Israel	Rothschild and Darr (2005)	Case study. Interviews with entrepreneurs, and incubator staff.	Social links (formal and informal).	Entrepreneurs argue that affiliation with the incubator gives a reputation and (scientific) credibility to the firm and hence helps firms to access business networks and more importantly funding.
Brazil	Etzkowitz, Mello, and Almeida (2005)	Case study based on detailed interviews.		The most important finding is that incubators allowed Brazil to create a less costly development strategy that took advantage of resources, such as academic, available elsewhere.
Finland	Totterman and Sten (2005)	Case study (3 incubator managers, 9 tenants, and 9 post-incubated tenants).		Incubator support and networking is important for firms to benefit from incubator resources. Incubator managers should focus on strategic business networking rather than only providing infrastructure and physical capital to entrepreneurs.
USA	Rothaermel and Thursby (2005a)	Panel of 6 years, 79 firms in Advanced Tech. Dev. Centre and Georgia Inst. of Tech.	Firm survival.	The effect of a university link (licensed technology, faculty as senior management, informal links) reduces the likelihood of failure but increases the incubation period. These effects become stronger if the inventor is the founder of the firm and/or takes part in the management of the firm.
USA	Rothaermel and Thursby (2005b)	Panel of 6 years, 79 firms in Advanced Tech. Dev. Centre and Georgia Inst. of Tech.	Revenue, total venture capital funding obtained, and firm survival.	Two mechanisms are investigated: (i) transfer by a licence to university, (ii) backward citations of incubator firm patents to university patents or publications. Holding a licence is important for firm survival but has no effect on other performance indicators. The absorptive capacity, measured by backward citations, is more important in explaining firm performance.
UK	Wynarczyk and Raine (2005)	Surveys in 17 incubators. Quantitative and qualitative evaluation.		Incubators do play an important role in nurturing business and creating jobs. The hands on support provided by the incubator and advisors are found to be vital for firm survival especially in the early stages of the business.
Italy	von Zedwitz and Grimaldi (2006)	Case study of 15 incubators. Qualitative assessment.	Incubator services provided.	Incubator services should be type-specific and the portfolio of the services provided should match with the objectives of the incubator. The incubator management should be sufficiently experienced to match service needs to incubator purpose.
South Korea	Kim and Armes (2006)	Data on 150 incubator managers (RR 40%). Qualitative assessment.		Qualified incubator managers tend to provide better services, use a wider range of support services, and establish wider networks. The rapid growth of incubators created a shortage of managers that in turn affects success.

Continued

Appendix table 10.1 continued

Country	Study (year)	Sample and methodology	Indicator for success	Findings
USA and Finland	Studdard (2006)	Survey of 52 firms. RR 18%. Quantitative analysis.	New products, techn. competence, reputation.	Knowledge acquired by interacting with the incubator manager has no effect on new product development and technological competence but enhances firm reputation (increased credibility, marketing reference).
USA and Europe	Gassmann and Becker (2006)	47 interviews and 77 firms from the EC benchmarking survey		The incubator and tenants benefit from resource and information flows at the initial phase. In the second phase benefits arise from intangible and tacit knowledge coming from the for-profit incubators and the firms.
China	Chandra, He, and Fealey (2007)	Case study of 12 incubators. Interviews with managers.		The fact that most incubators are large in size, high-tech-oriented, and rely on government funding made incubators dependent on government and weakened their capability toward market-oriented incubation.
Israel	Avnimelech, Schwartz, and Bar-El (2007)	Firm level quantitative assessment (3747 firms). Descriptive analysis.	Firm survival.	Venture capital (VC) and incubator programme are complementary. VC firms' failure rate is lower than incubator firms' failure rate. Closure rates are lower for firms that were established in an incubator and received seed funding (19% compared to 36%).
Europe	Aerts, Matthyssens, and Vandenbempt (2007)	Data from European Commission benchmarking study. 107 incubators. RR 18%.	Survival rates.	Tenants' survival rate is positively correlated with the availability of a more balanced screening process. Reliance on one screening process (market, financial, management screening) is positively related to high failure rate. Incubators' role in supporting entrepreneurial spirit by any means is critical for firm survival.
Finland	Hytti and Maki (2007)	131 high-tech firms (average RR 83%).		Firms that are young and have growth potential benefit more from the incubator services. Older firms tend to be less satisfied with services. Incubation period should be optimal and flexible according to firm needs.
UK	Hughes, Ireland, and Morgan (2007)	Interviews with 211 incubator firms. RR 25%. Cluster analysis to classify firms.	Product innovation, radical innovation, competitive performance.	Firms are classified into four groups according to their capabilities, access resources, and to acquire knowledge. Firms that deliberately network are more successful than others. Strategic networking is a crucial factor behind success and in its absence incubator services may not be important at all.

Ireland	McAdam and Marlow (2007)	Case study of one university incubator. Detailed interviews with 12 entrepreneurs.		Incubator facilities, its credible status, and networking opportunities are found to be important for tenants. However, firms were hesitant to share sensitive information and being in close proximity to each other might have adverse affects (secrecy, copying idea, etc.). Trust is a critical factor that enables information exchange.
Turkey	Akçomak and Taymaz (2007)	Matched sample of 48 incubator firms (RR 60%).	Sales, employment growth, innovativeness.	There are differences between on- and off-incubator firms in terms of sales and employment but not in innovativeness. Tangible incubator services and seed-funding explain this differential.
Ireland and UK	McAdam and McAdam (2008)	18 university incubator firms over 36 months.		Tangible incubator resources are important for the development of the firm in the early stages. Among a set of factors networking and clustering are rated to be the most important factor behind firm success.
Germany	Schwartz and Hornych (2008)	37 expert interviews in media incubators. Descriptive.		Availability of specialized equipment and sector-specific knowledge are crucial for the survival of media firms. Networking is constrained in the incubator because companies compete in the same sector.
Israel	Frenkel, Shefer, and Miller (2008)	12 incubators (6 private and 6 public). Surveys with incubator managers, 60 firms.		Technology incubators promoted technological entrepreneurship among the immigrants from US and former USSR. Firms in private incubators seem to benefit from networking with (international) strategic partners and academia. But private incubators cannot fully substitute public incubators.

Appendix table 10.2 Comparisons between developing countries regarding incubator policies

	BRAZIL	CHINA	INDIA	TURKEY	USA
Number of incubators	400	500	50	40	1000
Strategic focus	Mixed. Entrepreneurship, reduce unemployment, transfer of technology	High-tech focus. Foster entrepreneurship	High-tech focus but also traditional incubators to create ventures and jobs	High-tech focus but also traditional incubators to create ventures and jobs	Mixed. Transfer of technology, economic development
Emergence and evolution	Bottom-up	Top-down	Top-down	Top-down	Bottom-up
Scale	Small (average 15–20 firms per incubator)	Big (average 60 firms per incubator)	Small and smaller (<10 firms per incubator)	Small (average 15–20 firms per incubator)	Small and some larger incubators
Incubator funding	Government, business, universities	Government	Government	Government	Government, universities, for-profit business funds
Incubator services	Tangible and (poor) intangible	Mostly tangible and poor intangible	Tangible and (poor) intangible	Tangible and (poor) intangible	Tangible and intangible specialized services
Tenant entry/exit	Clear criteria and implementation	No clear criteria (especially exit criteria)	——	Entry criteria good but exit poorly implemented	Clear criteria and implementation
Incubator management	Strong	Poor	Poor	Modest	Strong
Role of government	Modest level	Very active	Present	Active	Low rather supportive
Role of university	Very active	Present	Present	Active	Very active
Institutional environment	Developing	Weak but developing	Weak but developing	Developing	Developed
Culture	Risk-averse	Risk-averse	Risk-averse	Modest risk-taking	Risk-taking

Note: Partially adapted from Chandra (2007).

Notes

1. The market failure argument justifies the need for incubators. However, incubators can be driven by other theoretical approaches such as the real options view and the resource based view. For a summary of these different approaches, see Hackett and Dilts (2004a).
2. This is also what differentiates incubators from business development services such as training, capacity building, fund raising, etc. Business development services could be provided by any company or institution for profit or not for profit. However, providing these services under one roof is a tool for incubators to achieve new venture and technology creation and create synergies.
3. There are three other factors that played a role in this trend as suggested by Hackett and Dilts (2004b): (i) the Bayh–Dole Act in the US that reduced the risk and uncertainty of commercializing publicly funded research, (ii) IPR system, and (iii) the commercialization of bio-medical research. The collapse of the dot.com market slowed this trend and wiped out virtual incubators. However, the general trend in the industry was not affected by this development.
4. For instance, in the UK, £50 million are available yearly via the Government National Innovation Fund. However, how much exactly is spent on incubators is not clearly documented. In a recent speech US president Obama declared that US $250 million will be invested in business incubation each year.
5. It is known that UNDP has supported incubator projects in several countries, including Turkey, Romania, Egypt, Uzbekistan, Colombia, and Pakistan. Regional development banks such as the Asian Development Bank also supported incubation projects, for instance in the Kyrgyz Republic.
6. For more information about *info*Dev, see <http://www.infoDev.org/business incubation>.
7. ANPROTEC webpage <http://www.anprotec.org.br/>
8. From the presentation of R. M. P. Jawahar, Executive Director of Triuchirappalli Regional Engineering College—Science and Technology Entrepreneurs Park (TREC—STEP). Available at <http://www.infodev.org/en/Publication.34.html>.
9. The study was an attempt to evaluate the early phase of the Turkish experience in technology business incubation. Information on 48 on-incubator (60 per cent response rate) and 41 off-incubator firms was collected by face-to-face interviews to compare the firms that benefit from incubator services with those that do not.
10. For a good attempt in this direction see Kim and Ames (2006). Success factors and best practices reported in various studies conducted by the OECD, NBIA, researchers, and practitioners are summarized in this study.

References

Abetti, P. A. (2004) 'Government-supported Incubators in the Helsinki Region, Finland: Infrastructure, Results, and Best Practices', *Journal of Technology Transfer*, 29 (1): 19–40.

Adegbite, O. (2001) 'Business Incubators and Small Enterprise Development: The Nigerian Experience', *Small Business Economics*, 17 (3): 157–66.

Aernoudt, R. (2004) 'Incubators: Tool for Entrepreneurship?', *Small Business Economics*, 23 (2): 127–35.

Aerts, K., P. Matthyssens, and K. Vandenbempt (2007) 'Critical Role and Screening Practices of European Business Incubators', *Technovation*, 27 (5): 254–67.

Akçomak, İ. S., and E. Taymaz (2007) 'Assessing the Effectiveness of Incubators: The Case of Turkey'. In V. V. Ramani and A. V. Bala Krishna (eds), *Business Incubation: An Introduction*. Hyderabad: Icfai University Press.

Allen, D., and V. Levine (1986) *Nurturing Advanced Technology Enterprises: Emerging Issues in State and Local Economic Development Policy*. New York: Prager.

Allen, D. N., and R. McCluskey (1990) 'Structure, Policy, Services and Performance in the Business Incubator Industry', *Entrepreneurship, Theory and Practice*, 15 (2): 61–77.

Allen, D. N., and S. Rahman (1985) 'Small Business Incubators: A Positive Environment for Entrepreneurship', *Journal of Small Business Management*, 23: 12–22.

Almeida, M. (2005) 'The Evolution of the Incubator Movement in Brazil', *International Journal of Technology and Globalization*, 1 (2): 258–77.

Autio, E., and M. Klofsten (1998) 'A Comparative Study of Two European Business Incubators', *Journal of Small Business Management*, 36: 30–43.

Avnimelech, G., D. Schwartz, and R. Bar-El (2007) 'Entrepreneurial High-tech Cluster Development: Israel's Experience with Venture Capital and Technological Incubators', *European Planning Studies*, 15 (9): 1181–98.

Becker, B., and O. Gassmann (2006) 'Corporate Incubators: Industrial R&D and what Universities Can Learn from Them', *Journal of Technology Transfer*, 31 (4): 469–83.

Bollingtoft, A., and J. P. Ulhoi (2005) 'The Networked Business Incubator—Leveraging Entrepreneurial Agency?', *Journal of Business Venturing*, 20: 265–90.

Bowles, S., and H. Gintis (2002) 'Social Capital and Community Governance', *Economic Journal*, 112 (483): F419–36.

Bruton, G. D. (1998) 'Incubators as a Small Business Support in Russia: Contrast of University-Related U.S. Incubators with the Zelenograd Scientific and Technology Park', *Journal of Small Business Management*, 36: 91–4.

Centre for Strategy and Evaluation Services (2002) *Benchmarking of Business Incubators*. Kent: Centre for Strategy and Evaluation Services.

Chan, K. F., and T. Lau (2005) 'Assessing Technology Incubator Programs in the Science Park: The Good, the Bad and the Ugly', *Technovation*, 25: 1215–28.

Chandra, A. (2007) 'Approaches to Business Incubation: A Comparative Study of the United States, China and Brazil', Working Paper 2007-WP-29. Indianapolis: Networks Financial Institute.

Chandra, A., W. He, and T. Fealey (2007) 'Business Incubators in China: A Financial Services Perspective', *Asia Pacific Business Review*, 13 (1): 79–94.

Colombo, M. G., and M. Delmastro (2002) 'How Effective Are Technology Business Incubators: Evidence from Italy', *Research Policy*, 31: 1103–22.

Cotis, J.-P. (2007) 'Entrepreneurship as an Engine for Growth: Evidence and Policy Challenges', paper presented at the GEM Forum: Entrepreneurship: Setting the Development Agenda, London, UK, 10–11 January. Available at <http://www.oecd.org/dataoecd/4/51/38031895.pdf>.

Etzkowitz, H., J. M. Carvalho de Mello, and M. Almeida (2005) 'Towards "Meta-innovation" in Brazil: The Evolution of the Incubator and the Emergence of a Triple Helix', *Research Policy*, 34 (4): 411–24.

Frenkel, A., D. Shefer, and M. Miller (2008) 'Public versus Private Technological Incubator Programmes: Privatizing the Technological Incubators in Israel', *European Planning Studies*, 16 (2): 189–210.

Fry, F. L. (1987) 'The Role of Incubators in Small Business Planning', *American Journal of Small Business*, 12 (1): 51–62.

Gassmann, O., and B. Becker (2006) 'Towards a Resource-based View of Corporate Incubators', *International Journal of Innovation Management*, 10 (1): 19–45.

Grimaldi, R., and A. Grandi (2005) 'Business Incubators and New Venture Creation: An Assessment of Incubating Models', *Technovation*, 25 (2): 111–21.

Hackett, S. M., and D. M. Dilts (2004a) 'A Real Options-driven Theory of Business Incubation', *Journal of Technology Transfer*, 29 (1): 41–54.

——— ——— (2004b) 'A Systematic Review of Business Incubation Research', *Journal of Technology Transfer*, 29 (1): 55–82.

Hansen, M. T., H. W. Chesbrough, N. Nohria, and D. Sull (2000) 'Networked Incubators. Hothouses of the New Economy', *Harvard Business Review*, September–October: 74–84.

Harwit, E. (2002) 'High-technology Incubators: Fuel for China's New Entrepreneurship?', *The China Business Review*, 29 (4): 26–9.

Hsu, P. H., J. Z. Shyu, H. C. Yu, C. C. You, and T. S. Lo (2003) 'Exploring the Interaction between Incubators and Industrial Clusters: The Case of the ITRI Incubator in Taiwan', *R&D Management*, 33 (1): 79–90.

Hughes, M., R. D. Ireland, and R. E. Morgan (2007) 'Stimulating Dynamic Value: Social Capital and Business Incubation as Pathway to Competitive Success', *Long Range Planning*, 40: 154–77.

Hytti, U., and K. Maki (2007) 'Which Firms Benefit Most from the Incubators', *International Journal of Entrepreneurship and Innovation Management*, 7 (6): 506–23.

*Info*DEV (2008) 'Financing Technology and Entrepreneurs & SMEs in Developing Countries: Challenges and Opportunities. India Country Study', Internet Source. Accessed on 12.09.2008 at <http://www.infodev.org/en/Publication.545.html>.

Jäckle, A. E., and C. A. Li (2006) 'Firm Dynamics and Institutional Participation: A Case Study on Informality of Micro Enterprises in Peru', *Economic Development and Cultural Change*, 54 (3): 557–78.

Jawahar, R. M. P. (no date) 'Developing ICT Synenergies for Incubating Start-Ups'. Triuchirappalli Regional Engineering College & Science and Technology Entrepreneurs Park (TREC-STEP). Available at <http://www.infodev.org/en/Publication.34.html>.

Kim, H., and M. Ames (2006) 'Business Incubators as Economic Development Tools: Rethinking Models Based on the Korea Experience', *International Journal of Technology Management*, 33 (1): 1–24.

Lalkaka, R. (2002) 'Technology Business Incubators to Help Build an Innovation-based Economy', *Journal of Change Management*, 3 (2): 167–76.

Lalkaka, R., and J. Bishop (1996) 'Business Incubators in Economic Development', *Journal of Change Management*, 3 (2): 167–76.

Leblebici, H., and N. Shah (2004) 'The Birth, Transformation and Regeneration of Business Incubators as New Organisational Forms: Understanding the Interplay between Organisational History and Organisational Theory', *Business History*, 46 (3): 353–80.

Lee, S. S., and J. S. Osteryoung (2004) 'A Comparison of Critical Success Factors for Effective Operations of University Business Incubators in the United States and Korea', *Journal of Small Business Management*, 42 (4): 418–26.

Ling, F. M., Y. Zhen-Jun, D. Gui-Lan, H. Bao-Fu, L. Shao-Hua, W. Xiao-Lin, S. Da-Hai, C. Ming-Xuan, L. Xiang, L. Nan-Lin, and W. Kun (2007) 'Promoting Business and Technology Incubation for Improved Competitiveness of Small and Medium-sized Industries Through Application of Modern and Efficient Technologies in China'. In V. V. Ramani and A. V. Bala Krishna (eds), *Business Incubation: An Introduction*. Hyderabad: Icfai University Press.

Löfsten, H., and P. Lindelöf (2002) 'Science Parks and Growth of New Technology Based Firms: Academic–Industry Links, Innovation and Markets', *Research Policy*, 31: 859–76.

Lundin, N., F. Sjöholm, and J. Qian (2006) 'The Role of Small Firms in China's Technology Development', Working Papers Series 227. Stockholm: European Institute for Japanese Studies.

McAdam, M., and S. Marlow (2007) 'Building Futures or Stealing Secrets? Entrepreneurial Cooperation and Conflict within Business Incubators', *International Small Business Journal*, 25: 361–82.

McAdam, M., and R. McAdam (2008) 'High Tech Start-ups in University Science Park Incubators: The Relationship between the Start-up's Lifecycle Progression and Use of the Incubator's Resources', *Technovation*, 28 (5): 277–90.

McAdam, M., B. Galbraith, R. McAdam, and P. Humphreys (2006) 'Business Processes and Networks in University Incubators: A Review and Research Agenda', *Technology Analysis and Strategic Management*, 18 (5): 451–72.

Mian, S. A. (1994) 'Are University Technology Incubators Providing a Milieu For Technology-based Entrepreneurship?', *Technology Management*, 1: 86–93.

——— (1996a) 'Assessing the Value-added Contributions of University Technology Business incubators to Tenant Firms', *Research Policy*, 25: 325–35.

——— (1996b) 'The University Business Incubator: a Strategy For Developing New Research/Technology-based Firms', *Journal of High Technology Management Research*, 7: 191–208.

——— (1997) 'Assessing and Managing the University Technology Business Incubator: An Integrative Framework', *Journal of Business Venturing*, 12: 251–85.

Miller, R., and M. Cote (1987) *Growing the Next Silicon Valley*. Lexington, MA: Lexington Books.

Nowak, M. J., and C. E. Grantham (2000) 'The Virtual Incubator: Managing Human Capital in the Software Industry', *Research Policy*, 29 (2): 125–34.

Oliveira, J. B., and A. C. M. Menck (2008) 'Model for Incubator Success: A Case Study in Brazil'. Presentation at XXV IASP World Conference, Johannesburg, South Africa, 14–17 September.

O'Neal, T. (2005) 'Evolving a Successful University-based Incubator: Lessons Learned From the UCF Technology Incubator', *Engineering Management Journal*, 17 (3): 1125.

Pena, I. (2004) 'Business Incubation Centers and New Firm Growth in the Basque Country', *Small Business Economics*, 22 (3–4): 223–36.

Peters, L., M. Rice, and M. Sundararajan (2004) 'The Role of Incubators in the Entrepreneurial Process', *Journal of Technology Transfer*, 29 (1): 83–91.

Phillips, R. G. (2002) 'Technology Business Incubators: How Effective as Technology Transfer Mechanism?', *Technology in Society*, 24: 299–316.

Rodrik, D. (2008) 'Second-best Institutions', Working Paper 14050. Cambridge, MA: National Bureau of Economic Research.

Roper, S. (1999) 'Israel's Technology Incubators: Repeatable Success or Costly Failures', *Regional Studies*, 33 (2): 175–80.

Rothaermel, F. T., and M. Thursby (2005a) 'Incubator Firm Failure or Graduation? The Role of University Linkages', *Research Policy*, 34 (7): 1076–90.

——— ——— (2005b) 'University-incubator Firm Knowledge Flows: Assessing Their Impact on Incubator Firm Performance', *Research Policy*, 34 (3): 305–20.

Rothschild, L., and A. Darr (2005) 'Technological Incubators and the Social Construction of Innovation Networks: an Israeli Case Study', *Technovation*, 25: 59–67.

Saravanan, R. (2007) 'Technology and Business Incubation in India: Challenges and Opportunities'. In V. V. Ramani and A. V. Bala Krishna (eds), *Business Incubation: An Introduction*. Hyderabad: Icfai University Press.

Scaramuzzi, E. (2002) 'Incubators in Developing Countries'. *Info*Dev Program. Washington, DC: World Bank.

Schwartz, M., and C. Hornych (2008). 'Specialization as Strategy for Business Incubators: An Assessment of the Central German Multimedia Center', *Technovation*, 28: 436–49.

Shefer, D., and A. Frenkel (2002) 'An Evaluation of the Israeli Technological Incubator Program and its Projects'. Israeli Financing Instruments for the Support of Entrepreneurship.

Similor, R., and M. Gill (1986) *The New Business Incubator: Linking Talent, Technology and Know-How*. Lexington: Lexington Books.

Studdard, N. L. (2006) 'The Effectiveness of Entrepreneurial Firms' Knowledge Acquisition from a Business Incubator', *International Entrepreneurship and Management Journal*, 2: 211–25.

Sweeney, G. P. (1987) *Innovation, Entrepreneurship and Regional Development*. London: Francis Pinter.

Tamasy, C. (2007) 'Rethinking Technology-oriented Business Incubators: Developing a Robust Policy Instrument for Entrepreneurship, Innovation, and Regional Development?', *Growth and Change*, 38 (3): 460–73.

Thierstein, A., and B. Wilhelm (2001) 'Incubator, Technology and Innovation Centres in Switzerland: Features and Policy Implications', *Entrepreneurship and Regional Development*, 13 (4): 315–31.

Totterman, H., and J. Sten (2005) 'Start-ups: Business Incubation and Social Capital', *International Journal of Small Business*, 23: 487–511.

Von Zedtwitz, M, and R. Grimaldi (2006) 'Are Service Profiles Incubator-specific? Results from an Empirical Investigation in Italy', *Journal of Technology Transfer*, 31 (4): 459–68.

Wennekers, S., and A. R. Thurik (1999) 'Linking Entrepreneurship and Economic Growth', *Small Business Economics*, 13: 27–55.

Wynarczyk, P., and A. Raine (2005) 'The Performance of Business Incubators and Their Potential Development in the North East Region of England', *Local Economy*, 20 (2): 205–20.

Yunos, M. G. M. (2002). 'Building an Innovation-based Economy: The Malaysian Technology Business Incubator Experience'. *Journal of Change Management*, 3 (2): 177–88.

11

Overcoming Adversity in Entrepreneurship-led Growth:

Evidence from the Indian Software Sector

Suma Athreye

11.1 Introduction

In many developing countries of the world the conditions for large-scale effective business development through entrepreneurial activity are far from optimal. Research by de Soto (2000) and Djankov and McLiesh (2005) and others shows that starting and running a business is extremely difficult in many such countries. Taking this argument further, Djankov and McLiesh (2005) say that a reform of the business environment could have a positive impact on the growth of some of the poorest countries. They contend that there is a positive relationship between the 'ease of doing business' and the human development index.

The problems posed by an adverse business environment can go much deeper than the mere inconvenience and costs of delays caused by regulation. As Krueger (1974) and Baumol (1990) argue, excessive government regulation and intervention often function as a means of rent extraction by particular groups in society. For example, the granting of licences by government officials frequently leads to competition for large rents, encouraging bribery, and diverting entrepreneurs into rent-seeking and away from innovative activities. In extreme cases, the perception that businesses become successful by exerting influence or bribing officials 'to do what they ought in any event to do' undermines both the link between pecuniary reward and business efficiency and trust in the motives and actions of governments (Krueger 1974).

Favouritism towards certain business groups can lead to the perception that government policy is a mechanism for rewarding the already rich and influential. This erodes values for doing business legally and ethically.

In India, the problem of an adverse business climate is especially acute. India is ranked 116th (out of 155 countries) in a ranking which shows how difficult it is to do business according to a series of criteria (World Bank 2006). The country also ranks 130th in terms of difficulties in trading across borders and 138th for the ease in the enforcing of contracts. Senior managers in India spent 12.9 per cent of their time dealing with requirements of regulations compared with 6.4 per cent average worldwide. Indian officials' interpretations of regulations are highly inconsistent and licensing laws in India have been notoriously difficult to navigate.

Nevertheless, despite the many difficulties of an adverse business environment, in recent years we have seen an explosion of technology-based entrepreneurship in India's software, information technology (IT), and business process outsourcing (BPO) industries. How can we explain this? The kind of techno-entrepreneurship witnessed in India faces numerous institutional constraints, only some of which are imposed by or presided over by the government. Serious constraints arise from underdeveloped financial markets, poor protection for property rights, and weak contract enforcement. One would expect these constraints to erode profitability, restrain market entry, and impose high transaction costs on new entrepreneurial ventures, thereby stifling creativity and innovation. In theory, the Indian software industry 'should not' have developed in the way it did.

This chapter shows how the Indian software industry achieved its astonishing results despite the adverse conditions facing entrepreneurs.[1] When given the economic opportunity, domestic entrepreneurs developed new world class business models, and provided a demonstration effect for other industries to follow. In turn, these changes created the conditions for more widespread institutional transformation and reform of the business environment in the economy.

Technology-based entrepreneurship in the Indian software sector thus provides vital lessons for our understanding of the constraints to entrepreneurship in other countries with poor business environments.

11.2 Technology-based entrepreneurship and its role in development

11.2.1 *Entrepreneurship and the economic environment*

Entrepreneurship studies can be traced back to the work of Richard Cantillon (circa 1730) and Jean Baptiste Say (1816).[2] Cantillon saw entrepreneurs as actors who are typically able to cope with uncertainty, while Say (1816) saw the entrepreneur as the agent who combined all means of production in order to make profits. These ideas about what entrepreneurs did were rediscovered in the twentieth century. Thus, Frank Knight (1921) emphasized the entrepreneur's role in coping with the uncertainty of market dynamics, arguing that entrepreneurs were also required to perform fundamental managerial functions such as direction and control. Harvey Leibenstein in the 1960s and 1970s saw the entrepreneur as the agent who resolved market deficiencies through input completing activities. A somewhat different twist to the advantages of entrepreneurship was given by Joseph Schumpeter (1934). He saw the entrepreneur as a heroic innovator who implements change within markets through the realization of 'new combinations' of various kinds. For Israel Kirzner (1979) the entrepreneur is the one who recognizes and acts upon market opportunities. More recent contributions by Rothwell and Zegveld (1982) identified 'intracorporate' entrepreneurship (or intrapreneurship) in the modern context, showing how managers can create new businesses within large corporations. None of these writers explicitly examine entrepreneurship within the context of the developing countries or ask how the modern problems of development described by de Soto (2000) affect the ability of entrepreneurs to operate. These include, as we noted earlier, the presence of underdeveloped financial markets, poor protection of property rights and weak contract enforcement. Most entrepreneurship research today implicitly assumes that there is no difference between the entrepreneurship being carried out in the most developed nations and that carried out in 'latecomer' or developing countries.

Similarly, a surprising aspect of the development literature that developed through the 1950s and 1960s is that it never considered entrepreneurship as a serious agent of development in poor economies, despite the important role played by entrepreneurship in the overthrow of feudal economies in the First World. The first paper that looked at the issue of entrepreneurship and the development of poor economies was probably Leff (1974) who identified poor institutions as an impediment to entrepreneurship. Yet, it was not until the 1990s and the works of Baumol (1990, 2005) and North (1990, 2005) that significant theoretical insights emerged about the relationship between the quality of institutional environment and type of entrepreneurship. To some extent these theories still provide the important arguments about why entrepreneurship is not likely to flourish in developing countries. While Baumol argued that institutions enshrine incentive structures for rent-seeking as opposed to productive entrepreneurship that would contribute to growth, North identified the essential endogeneity of institutions and economic activity. Entrepreneurial firms will adapt their activities and strategies to exploit opportunities and overcome the limitations in their business environments through formal and informal institutional networks.

Two aspects of the formal institutional environment pose particular problems in developing economies: poor private property protection and poor capital market institutions. Much of our knowledge about the impact and responses to these two problems comes from studies of transition economies. One predominant effect of poor private property protection is corruption which increases the costs of doing business. Johnson, McMillan, and Woodruff (1999) argue that businesses in transition economies generally circumvent this problem by relational contracting (i.e. contracts informally enforced through networks). Access to informal networks to alleviate the shortcomings of the environment and the resources of the firm is also seen as relevant in developed market economies, where entrepreneurs with good ideas may nonetheless lack all the resources required for successful new firm formation. Johnson, McMillan, and Woodruff (2002) also argue that reputational incentives often substitute for legal enforcement of contracts. What these studies show is that though poor institutional environments provide obstacles which may often reduce the profitability of an economic opportunity, entrepreneurial firms can and do devise ways to get around them.

The absence of capital market institutions and other intermediate markets for industrial goods and services can hinder the emergence of entrepreneurial businesses because they make the risk of starting a business quite high. The role of these institutions in facilitating the formation of business groups in emerging economies has been stressed by Khanna and Palepu (2000). In the context of capital scarcity and the large initial scale of production required to compensate for missing intermediate markets, business groups are able to derive the advantages of risk pooling and cross-subsidization in order to overcome problems of lack of availability of finance. This feature probably also makes them more likely to undertake greater risks. This is of course very relevant for technology-based entrepreneurship which is likely to suffer particularly as a result of high riskiness and information asymmetries with regard to capital borrowing. In developing economies thus, we should therefore expect business groups to play an important role in technology-based entrepreneurship and the evidence of the East Asian and South Asian experience certainly confirms this reasoning.[3]

11.2.2 *Incumbent firms and the barriers to technology-based entrepreneurship*

Shallow capital markets and poor intellectual property (IP) protection and enforcement may exert a significant influence upon the kind of business models chosen by technology entrepreneurs in developing countries. However, other constraints stem from the shallow nature of domestic demand in poor markets and the reactions of incumbent firms.

Technology products often face adverse demand conditions in developing countries. The relatively low income of consumers and firms often favours goods and services with low prices to goods of higher quality. Put differently, technology-based products do not find ready markets. Hobday and Perini (2005) note that 'latecomer' entrepreneurial functions differ in many ways from the conventional advanced country (or 'leadership') entrepreneurship which focuses on developing new products and technologies. In particular they point out that latecomer technology entrepreneurs are more focused on delivering incremental improvements in goods and services exported to foreign firms and transnational corporations, rather than to a large domestic market. Why should this be the case?

Even where such markets exist the successful exploitation of innovations needs much more than the capability to produce innovations and the funds to invest in R&D. One reason for this is the presence of incumbent firms who can imitate most innovations unless very strong intellectual property rights (IPRs) protect inventors. In a seminal paper, Teece (1986) highlighted the role of environmental conditions that influenced appropriability and the control over complementary assets required to commercialize a technology as the two principal factors that influence how much a firm will profit from an innovation-based strategy.

In particular, Teece (1986) argued that successful innovation often needed other complementary assets such as brand value, marketing networks, control of distribution channels, etc. which are often possessed by incumbent firms. Weak appropriability conditions such as those associated with weak IPRs, when combined with control over complementary assets such as marketing and distribution, will favour vertical integration by incumbents as the dominant strategy for the commercialization of innovation. In such situations, large incumbent firms, with their control of large distribution networks, are best positioned to profit from technology commercialization strategies. In developing country contexts, entrants into the technology space with deep pockets may be able to invest in such complementary assets. Thus strong capital market positions may be favourable to technology entrepreneurship. However, entrants may often be forced into situations where they have to negotiate with the dominant firm, especially if they cannot fully appropriate the benefits of technology they have developed, which is likely to be the case with small incremental innovations. In contrast, strong appropriability conditions (such as provided by tight IPRs) would favour business models based on licensing of technology by new and smaller entrants. Thus, in some markets 'first movers' and incumbents prevail in the market with certain innovations, while in other situations 'followers' and later entrants are able to gain the lion's share of the profits. To quote from Teece (1986: 285):

> when imitation is easy, markets don't work well, and the profits from innovation may accrue to the owners of certain complementary assets, rather than to the developers of the intellectual property. This speaks to the need, in certain cases, for the innovating firm to establish a prior position in these complementary assets... innovators with new products and processes which provide value to consumers may sometimes be so ill positioned in the market that they necessarily will fail.

Thus, the management literature on technology entrepreneurship suggests that in understanding cases of successful and unsuccessful technology entrepreneurship in developing economies, one needs to look beyond just the environmental conditions and the technological capabilities of firms. In oligopolistic markets, incumbents may be able to imitate many technological developments easily because of their strengths in the possession of complementary assets.[4] Battling such firms can be hard work for technology entrepreneurs unless they have a very radical innovation (based on a different competence set) that is effectively protected by IPRs. Here too the Indian software industry provides a good example. Although more recent efforts at entrepreneurship have tried to better the outsourcing model by exploiting some niches in the semi-conductor market, by and large Indian software firms have avoided the head-on competition with large incumbent multinational enterprises (MNEs). One may argue whether this reflects timidity of ambition or a wise appreciation of the relative power of incumbent firms, but cooperating with the incumbents has certainly enhanced the longevity of software service firms.

11.2.3 *Policies to promote entrepreneurship*

Since Baumol's work on productive and unproductive entrepreneurship the reform of the institutional environment is seen as essential to fostering productive entrepreneurship. However, institutional reform, though easy to prescribe, is far more difficult to achieve. Although many developing countries have instituted laws on private property, their enforcement remains limited—partly due to the prevalence of different customary norms and partly due to poor judicial systems. As a result, the quality of many potentially growth-enhancing institutions remains poor and whilst many policymakers do recognize that growth would be stronger in the presence of better enforced property regimes, they are often unable to implement the required reforms that would make the quality of their institutions better. Bardhan (2006) draws on the Indian experience to argue that political impediments often come in the way of institutional change in the form of vested interests, distributive conflicts, and collective action on the part of disenfranchised groups in response to the proposed reforms. Put differently, it is unclear who should be the agents of institutional reforms even if institutional reforms were accepted as necessary to improve entrepreneurship. Perhaps in silent acknowledgement of this issue, policies to improve entrepreneurial performance in developing countries have often taken the form of the provision of finance to small-scale industries.

Yet whether or not special policies are needed to call forth more entrepreneurship in the economy is an area of considerable debate. An optimistic view of the importance and necessity of entrepreneurship in developing economies comes from Hausman and Rodrik (2003). They point out that in the presence of uncertainty about what a country is good at producing, there can be great social value to discovering cost advantages of domestic activities, because such discoveries can be easily imitated. However, as in the case of all externalities this would make entrepreneurship subject to underinvestment. Government policy should thus intervene to induce investment in entrepreneurship, while at the same time rationalizing excessive diversification once such investment has occurred.

In their own words:

> Neither economic theory nor management science is of much help in helping entrepreneurs (or the state) choose appropriate investments among the full range of modern sector activities, of which there could be tens of thousands, once one moves beyond broad categories such as 'labour-intensive products' or 'natural resource-based products'. Yet making the right investment decisions is the key to future growth, as it determines the pattern of specialization. In these circumstances, there is great social value to discovering that cut flowers, soccer balls, or computer software can be produced at low-cost, because this knowledge can orient the investments of other entrepreneurs. But the initial entrepreneur who makes the 'discovery' can capture only a small part of the social value that this knowledge generates... Consequently, entrepreneurship of this type—learning what can be produced—will typically be undersupplied, and economic transformation delayed. (Hausman and Rodrik 2003: 4)

This view stands in sharp contrast to Kirzner (2000) who first emphasized the discovery function of entrepreneurship.

> To the extent that [government policies] suspend or inhibit the market process, they are obstructing a process of discovery without offering any substitute for it. Let us not forget that the market process has the function of alerting market participants to opportunities which nobody has expected. To initiate governmental policies to grapple with externalities is, in effect, to pretend knowledge which no one can, in principle, honestly claim to possess. (Kirzner 2000: 82)

In the following sections we test the ideas reviewed in this section against the facts about the growth of the Indian software sector.

11.3 How Indian software firms overcame obstacles in the business environment

11.3.1 *The mediating role of the dominant business model*

Software outsourcing from India grew through the 1980s and 1990s despite weak laws to enforce contracts, poor IP protection, inadequate capital markets, and a policy regime that was generally market unfriendly. Software services exports from India have grown from a mere US$330 million in 1993 to US $17.3 billion in 2006, employing around 878,000 people.[5] Estimates in Athreye (2005) suggest that in 2001, entrepreneurial firms accounted for about 38 per cent of sales revenues and 35 per cent of employment in the Indian software outsourcing sector. Six of the top 20 firms were entrepreneurial in origin. The techno-entrepreneurship that sustained the growth of the Indian software industry is thus both intriguing and inspiring for other countries that are poor, face adverse regulations and institutions.

The simple answer to 'how did this happen?' is that the industry exploited its initial advantage in low-cost human capital by fashioning business models that leveraged this strength. As the newly emerging global IT industry boomed in the West, this led to a huge demand for trained engineers and technicians. Indian firms saw this economic opportunity and exploited their cost advantage by occupying product market spaces and business models that avoided the penalties of their poor institutional environment and head-on competition with incumbent firms. This occurred initially through the development of customized software designed for foreign multinational corporations (MNCs) within a business outsourcing relationship, in a manner very similar to the original equipment manufacturer arrangements of East Asian latecomer entrepreneurs in Korea and Taiwan (Hobday 1995). In the 1970s and 1980s the hurdles imposed by government regulations and institutional difficulties meant it was far easier for Indian software firms to move teams of engineers abroad—a practice that is sometimes referred to as 'on-site services' or more derogatorily as 'body-shopping'. In this way, the IPRs always belonged to the client firm and they could monitor the programmes directly.

The pioneering firm in establishing this model was Tata Consultancy Services (TCS), a subsidiary of the business house of Tatas. Despite being a large business group with deep pockets and entering into an industry (software) that was already reasonably well understood, TCS nevertheless had to behave like a pioneering firm because the industry was new to India. As a firm, it had to invest on many fronts—train engineers to learn software management skills as well as software languages, respond to changing technologies, and negotiate with the government to obtain permission to explore the business opportunities that presented themselves. TCS understood well the problems of market creation. It chose outsourcing as a way to avoid addressing the issue of who would buy its services. In choosing this path, the company also recognized the potential that such contracts offered for technology upgrading and in 1975 partnered exclusively with Burroughs. In this respect, its strategy was similar to that of firms in the Four Dragon countries. However, in 1978 the takeover bid for TCS by Burroughs also showed the limits of this early model. TCS and numerous later entrants would find contractual ways of avoiding this problem and used client diversification as an important means to neutralize such take-over threats. However, survival came at a price. Client diversification also created obstacles to building up the firm's own domain knowledge.

Although the basic outsourced business services model was established in the 1970s, it has continued to evolve through the 1980s and 1990s. The first steps towards a new form of the model were adopted in the 1990s when the offshore content of on-site delivery began to slowly increase. Some firms such as TCS and Infosys gained reputations for timeliness and product quality and started attracting more work (Banerjee and Duflo 2000). They also began to be trusted to do the work in India where they could deliver the same quality of software at an even lower price. As prices of telecommunications access came down through the 1990s, the off-shoring of work became steadily cheaper and also utilized the large educated labour force better as firms could now practice a fine division of labour in software teams due to the much larger scale of such operations. These features started subtly altering the nature of the market.

Although the 1990s had brought in a wave of imitative entry from other Indian firms and foreign MNEs, factors like reputation and better process efficiency started creating some differentiation between producers. Data presented in Athreye (2005) show a marked separation of the top quartile from the median firm on measures such as employment size, turnover, and average labour productivity by 1997. There is also some evidence of increasing marginal returns to labour with an output elasticity between 1.2 and 1.4. Thus, although the 1990s was a decade of imitative entry there was an endogenous sorting of firms where those with better process management capability were also adopting more (profitable) offshore modes of services delivery.

With full financial liberalization of the economy, firms such as Infosys also realized other advantages of operating in a global market. Being an entrepreneurial firm rather than a business house subsidiary, Infosys was vulnerable to periods of capital scarcity and this vulnerability characterized its first years of growth. To keep the potential of deep pockets for the future, they were the first to leverage their reputation as suppliers to list on international markets such as NASDAQ and the NYSE and then use their reputation of being professional companies and good corporate citizens in order to attract new US customers. They also voluntarily adhered to some corporate norms of US firms such as use of more transparent accounting practices and the use of employee stock options to stem attrition of their staff. These were both moves that business house subsidiaries could not imitate easily and which further reinforced the reputation of Infosys for professionalism and good corporate citizenship.

As the larger software services firms started gaining market share, new entrants in the late 1990s explored another new variant of the offshore model, viz. application of the software services delivery model to administrative functions of companies (the so-called BPO). A more recent wave of entrants has also focused on R&D outsourcing in the telecom domain. Each of these small variations of the offshore delivery model has tried to utilize India's advantage of a large and educated labour force whilst at the same time making use of developments in technologies (such as growing telecommunications reach) and the advantages of the global market for such services that had emerged. Thus, far from stifling entrepreneurial activity, the penalties of the domestic environment led Indian firms to be inventive in devising new value propositions to overcome that adversity. Indeed the dominant model has focused entrepreneurial energies along a narrow path of proven success.

The business model of outsourced software services took nearly two decades to evolve fully and in this time several firms built up process capabilities complementary to the successful global delivery of software and other services. The search for new entrepreneurial opportunities explored the product space and incremental improvements were made to the outsourced service model which was very successful.

11.3.2 *The roads less travelled: capability gaps exposed by business model experimentation*

Not all the experimentation with business models was in the form of incremental improvements. The desire to have an Indian software product that would be a proud symbol of the nation's software process has been a wish of many politicians and business persons. Businesses have also been attracted to the much higher profitability of the software business model and through the history of the industry's growth there have been recurrent entrepreneurial efforts to uncover a viable product model. Though they represent the road less travelled, their moderate success also exposes capability gaps and institutional constraints that eroded value and increased risk.

The first known case of a software product from India was not surprisingly from TCS. TCS had always paid attention to tools in software writing, and in the early 1980s the organization saw an opportunity to sell some of those tools as products. During the interviews, a product called Case Packet was mentioned, which launched to good reviews in the technology business press and retailed for about US$200,000. The company depended upon distributors for its marketing, and it was soon taken over by Computer Associates and never heard of again. Though TCS proved very capable of making products, as one of their managers summed it up, 'TCS did not create the product market, and that was the subtle difference.'

Another wave of experimentation was in the mid-1980s, following on the heels of import liberalization which greatly increased the installed capacity of PCs in India. Firms such as Sonata and Mastek introduced products aimed at the domestic market for software. Sonata tried to develop software products while Mastek became the first company to use tools to speed up product development again for the domestic market. However, these product ventures largely failed—due to the small domestic market and lax IP laws. The lack of venture capital (VC) support also made the product model a very risky one for entrepreneurial firms to adopt due to the higher up-front investment implicit in the business model.

The late-1980s saw a renewed exploration of the product market but this time aimed at foreign clients rather than domestic ones and carried out by larger firms who had sufficient financial resources pockets for such sustained investment. The stimulus was the spread of distributed computing and the opportunity to write products previously constructed for mainframes around the new technology platform of LAN and interlinked PCs. Thus, Ramco (a business house subsidiary) tried to develop an Enterprise Resource Planning Product (ERP) and CITIL (a spin-off from the multinational subsidiary of Citibank, COSL) developed a financial product Flexcube around their customization of a pre-existing market product called *Cosmos*—a company they also acquired subsequently. COSL thus acquired the full IPRs for their package in a relatively short period of time.

Ramco's product Marshall was overtaken by solutions by Western companies like SAP and other ERP product producers. When questioned about the relatively poor performance of Marshall, managers at Ramco conceded that it probably took longer for the company to produce the basic software infrastructure required for a successful product, while having been in the market longer, SAP could just translate to another technology platform quite quickly. However, they did manage to acquire a reasonable number of installations which helped the launch of their later product Ramco Virtual Works. Flexcube, however, emerged as the most successful product developed from India, although its ownership has kept changing. Its success was the result of a shrewd strategy which targeted customers in poor countries and among smaller firms that were unlikely to go after existing mainframe solutions. These firms were also less likely to become competitors or imitators.

In interviews, both firms emphasized the huge up-front investments needed and the important role for marketing capabilities in developing successful software products for overseas clients. A domestic market where firms were not using such products made the job of designing products for customers they did not know much harder.

The adoption of tighter IPR norms in 1995 gave rise to another burst of experiments with software products in the area of embedded software. Many telecom firms like Texas Instruments, Motorola, and Nortel had located their research on digital signal processing chips in India and this gave rise to spin-off firms. Two prominent examples are Sasken (previously a joint venture with Nortel) and Ittiam. Both hold patents to their names and aim to earn revenues largely through licensing to large firms. Sasken combines outsourced R&D services with licensing and this gives the firm stable cash flows and the ability to operate on a larger scale. Ittiam operates a pure product model and has been blessed with a large tranche of VC financing from Silicon Valley venture capitalists. They had filed for 11 patents in the first four years of their existence and their chips are licensed to an impressive list of client firms.

The lure of the product model is still strong but thinking about it has evolved and in the process has exposed gaps in capabilities and in the environment.

11.3.3 *Institutional reform and adaptation: the effect of entrepreneurship on the economic environment*

The spectacular growth of the industry in the 1990s was also marked by an improvement in the institutional infrastructure surrounding the software outsourcing industry, which generally served to ease the constraints on the industry's further growth. These included capital and labour market reform, better access to finance, improved IPR protection, and contract enforcement.

Capital market institutions did not understand how to evaluate the finance needs of the emerging software industry. Infosys, India's most famous entrepreneurial firm, was refused a bank loan when it was set up in 1981 and had to borrow the start-up money from the wife of one of the founders. It was probably not the only one. Faced with a situation where bank finance was not readily available and VC was not forthcoming, software firms were conservative in their own cash flow calculations but experimented with *importing the use* of capital market institutions in the US. Many software firms voluntarily listed on stock exchanges in the US and in Europe with more stringent disclosure norms in order to raise money for investments and acquisitions. The compliance of some firms to international norms was a powerful force for improved corporate governance with the chairman of Infosys being involved in committees to promote these changes.

After 1993, the combined effects of liberalization and the success of the software industry drew US VC into India. Dossani and Kenney (2002) show that a significant portion of the sevenfold increase in funds from 1993–8 was accounted for by the entry of foreign investors after 1995, through investment arms of foreign banks, and venture firms that had raised capital abroad. Indian Silicon Valley entrepreneurs encouraged new business plans within India and exploited the new exit routes made possible by the international listings of Indian software firms on NASDAQ, NYSE, and the LSE. According to recent estimates from the Indian Venture Capital Association, domestic and foreign venture capitalists invested US$774 million in 2003 in India up from US$590 million in 2002 (Nair 2004; Basu 2005).

Training and the supply of human capital also improved. As the software industry grew in the late 1980s and early 1990s, labour markets for software programmers became tight due to global market expansion and fierce competition. In this period, scores of privately funded and organized educational and training institutions emerged to meet the demands for skilled labour, expanding supply beyond what could be produced through the state-funded educational establishments. Desai (2003) shows that while engineering capacity increased six-fold from 61,000 in 1987–8 to 341,000, the supply of graduates with IT degrees saw a ten-fold increase, from 25,000 in 1987–8 to 258,000 in 2002–3.

Privately financed training institutes such as NIIT and Aptech Ltd. sprung up to provide software training throughout the 1990s—a dramatic institutional departure in a country where reliance on publicly funded training institutions had been the norm. As Table 11.1 below shows, their incidence increased all over India, especially post-1995.

Intriguingly, all of these changes occurred *after* the software growth opportunity had been spotted by entrepreneurs with some initial success. India's software firms did not wait for institutional reform. On the contrary, software success caused the reform to take place.

11.3.4 *Creating business friendly policies*

A very important agent for institutional reform was the National Association of Software Service Companies (NASSCOM), the largest and most important business association for software services and now BPO. NASSCOM was set up

Table 11.1 Privately financed IT institutes in India, by region

Region Year	East	Central	North	West	South
1987–8	12	25	5	80	66
1995–6	13	42	26	80	76
2002–3	84	76	82	86	92

Source: Adapted from Arora and Gambardella (2004: figures 2 and 3) based upon data from the All India Council on Technical Education.

in 1988, with just 38 members who collectively accounted for 65.0 per cent of the industry revenues. Many of these members were small companies, and the total industry revenue in 1988 was a little over US$100 million. By 2003–4, the number of members had risen to more than 800, collectively accounting for about 95 per cent of the industry output of about US$21 billion.

NASSCOM operates as a collective body representing the interests of the software sector with functions of lobbying, advocacy, and public relations (Kapur 2002). NASSCOM has been extremely effective in lobbying for policies favourable to the industry's continued growth, in collective marketing at a time when Indian companies did not have an international reputation for delivering quality service, and in providing information on the industry for insiders and outsiders. Collective marketing involved organizing trade fairs and producing directories of firms and their areas of business for potential customers, bringing together demand and the eager small entrepreneurial firms.[6] The difficulties of 'selling' Indian software at the time cannot be overstated as a founder-member of NASSCOM told us:

> When I was out there in 1991, the country was bankrupt. We had three governments in one year, an assassination of a prime minister, and we were hawking our gold. You know, selling overseas was not a piece of cake . . . if I have to present ten slides, the first eight had to be to sell India and the ninth one would say we do have an IT industry in India and unless the guy bought those nine slides, your tenth one about your company was meaningless. Because who are you anyway? So we had to build up the [Indian] brand from day one.

The attitude of NASSCOM towards engagement with the government on policy issues is a dramatic break with past practice. Older industry associations, such as the Federation of Indian Chambers of Commerce and Industry (FICCI) and the Confederation of Indian Industry (CII), had a more hands-off approach to policy engagement. NASSCOM engaged the government with facts to ensure that business was given a free hand to take initiatives. This approach has made for better industrial policy and improved corporate governance. The NASSCOM lobbying model has been emulated in other fast growing sectors, notably biotechnology and automotive components, which speaks volumes for its effectiveness.

11.4 The benefits of entrepreneurship

11.4.1 *Economic benefits*

In many developing countries start-up entrepreneurs fail to make much headway in terms of employment growth, new technology adoption, and capital efficiency (von der Fehr 1995; Beck, Dermirgüç-Kunt, and Levine 2003;

Shadlen 2004). To assess this issue, we compared some quantitative measures of the business performance of software entrepreneurial start-up firms with those of business house subsidiaries and foreign MNCs (see Table 11.2). Table 11.2 is based upon a survey of 204 software firms (132 entrepreneurial firms, 27 business house subsidiaries, 45 foreign firms). It shows that while the starting size (indicated by the median number of employees after the first year of business) for entrepreneurial firms was only marginally larger than that of domestic business house subsidiaries, both started smaller than foreign firms. Entrepreneurial firms grew to an employment size almost as large as foreign firms. A similar picture emerges when we compare size by turnover. Entrepreneurial firms showed a slower rate of growth than the subsidiaries of large domestic firms but grew at a marginally higher rate compared to foreign firms.

Turning our attention to initial capital outlays we find that entrepreneurial firms started with very small initial capital outlays compared to both domestic firm subsidiaries and foreign firms, especially the former. This may be related to the scarcity of finance that entrepreneurial firms face when they start up due to the high cost of capital. If we look at the ratio of the median turnover to median capital outlays as a crude input–output measure, we find that entrepreneurial firms are the most efficient in their use of capital. This ratio is more than two and a half times that for foreign firms and over three and half times

Table 11.2 Economic impact of various software entrants, 2003

	New start-ups & spin-offs	Subsidiaries of domestic firms	Foreign firms & MNE subsidiaries	All firms
Number of firms	132	27	40	204
Median annual rate of growth of employment (per cent per annum)	30	42	26	29.7
Median number of employees at the end of first year of operations	15	12	22.5	18
Median number of employees in 2003	80	143	90	100
Revenues				
Median annual revenue in 2003 (in Rs. million)	80	268	100	90
Median revenues/firm age*	11.43	43.67	12.5	12.86
Equity				
Median equity (initial capital outlay in Rs. million)	3	37.5	10	
Ratio of median revenues to median start-up equity	26.67	7.15	10.00	18.00

Note: revenue divided by firm age is a proxy for firm growth.

Source: Own survey on firm origins, 2008.

that for subsidiaries of domestic firms. In a capital scarce economy, entrepreneurial firms appear to be doing a better job of conserving capital than non-entrepreneurial firms.

Thus, two surprising conclusions emerge from our survey: first, the employment growth of entrepreneurial firms was at least as good as that of MNCs, second, entrepreneurial firms are more efficient than both domestic and foreign firms in their utilization of capital for the most productive use. These findings should bring cheer to many capital scarce economies.

11.4.2 *Business model benefits: the propensity to experiment*

Our survey also compares the motivations of entrepreneurial firms entering the software business with business house subsidiaries and multinational firms (see Table 11.3). While the pursuit of a profitable business opportunity was the overriding motive for all firms entering the software sector, the desire for independence and the possibility of technological innovation all figured highly in motivating entrepreneurial entry. By contrast, subsidiaries of domestic firms were more concerned with diversifying into more profitable areas and the desire to earn foreign exchange through exports.

Overall entrepreneurial firms had more creative space and were better placed to pursue their individual visions. Entrepreneurial firms were also more likely to experiment with new business models.

Table 11.3 Three most highly rated motivations for entering the software business

Motivation	Number of times a motivation is considered important*
Entrepreneurial firms (N=132)	
Identified new business opportunity	100
Desire for independence	74
Stimulated by research possibilities and the desire to innovate technologically	70
Business house subsidiaries (N=27)	
Diversification into a more profitable growth opportunity	19
Earn foreign exchange through exports	13
Growing software needs of the parent company	11
Foreign firms (N=40)	
High quality of programmers	35
Lower cost of programmers	29
English as international business language	25

Note: * Number of times a motivation is considered to be either important or very important, on a five point scale running from very unimportant to very important

Source: Own survey of firm origins, 2003.

11.4.3 *Impact of software entrepreneurship on other sectors*

Each of the institutional developments described in the previous sections have impacted other sectors where the outsourcing business model was adopted. Capital market reforms, which began as part of a larger financial reform process, have gathered steam. The successful use of international capital markets by Indian software firms and the simultaneous listing of software firms on both the Indian and foreign stock exchanges have resulted in a realignment of disclosure rules and corporate governance procedures in the Indian capital markets.[7]

The emergence of third party BPO activities in India led by firms such as EXL Services, 24/7, Spectramind, Daksh e-Services, and Transworks, who all received VC funding for their seed capital, has impacted on manufacturing, health care, banking and financial services, pharmaceuticals, engineering, and textiles. As a result there seems to be a gradual convergence in India towards the US model of VC institutions, initiated and aided by the diaspora of technology entrepreneurs in India and their Silicon Valley partners.

The NASSCOM model of industry–government interaction has been adopted by new sectors relying on domestic entrepreneurship. Examples include ABLE, the Association of Biotechnology Led Enterprises, which represents the biotechnology industry, and the older Automotive Component Manufacturers Association, established in 1958. The latter's activities since 1994 resemble the NASSCOM model quite closely. Indeed, a visit to the web sites of these organizations shows that their strategies and information content are similar to that provided by NASSCOM. This institutional reform has created more visibility about the desirability of more reforms and the part this process could play in supporting entrepreneurial growth in other services and knowledge-based sectors.

11.5 Conclusions

Contrary to conventional wisdom, Indian software firms were able to circumvent government-led restrictions to growth and then lead institutional reform in India. Moreover, these firms are bringing about major changes in the way business is being carried out in many other sectors, producing a wide ripple effect which will, hopefully, continue and grow in the future. The impetus for institutional reform has not come from government, international institutions, or their advisors, but primarily from the business sector itself, reversing the way development analysts normally think about economic progress.

An interesting factor in India's software success has been the long period of sustained export demand which was somehow untouched by the institutional problems that faced Indian software firms. The shortage of programming talent in the Western economies was, to use statistical terminology, largely exogenous to the institutional constraints facing Indian software entrepreneurs. The long period over which this shortage has persisted has ensured that Indian software firms have been able to consolidate their toe-holds in the export market into more secure market shares and also collectively lobby for institutional reforms that improve their operating environment. The market economy in India also responded to the economic opportunities provided by this boom (e.g. the spontaneous growth of software training institutions in the private sector). Had the demand opportunity come mostly from within the domestic economy, the institutional constraints faced by Indian software entrepreneurs might have sparked off various political conflicts over profit sharing (e.g. between labour and entrepreneurs, between firms in the sectors buying software and firms selling them) that might have undermined future growth.

We find very little evidence that imitative excessive entry in the software sector has slowed entrepreneurship down as suggested by Hausman and Rodrik (2003). Instead the response to competitive pressures from imitative entry has been to encourage new entrants to search for business models that add to variety. At the same, incumbent entrepreneurial firms have been able to hold their own position by exploiting traditional economies of scale.

In the software sector, start-up entrepreneurs played a leading role in creating and disseminating new business models and changing restrictive institutional practices. Indian software firms saw the burgeoning demand opportunity in the international market place and imported new institutional norms from the advanced nations, especially the US, bringing about improved capital inflows and enhanced IP protection. The new entrepreneurs not only helped reform local institutions, but also began building new institutions and practices which are now diffusing to other industries. Through their strategies, Indian start-up firms are changing the way business is done in India. This process of reform was made possible by first tapping into the economic opportunity offered by the huge boom in demand for IT services in the world economy.

Survey evidence also shows that the new start-up firms outperformed the advanced multinational corporations and the large local business house subsidiaries in terms of capital efficiency and were at least as effective in creating employment and developing the new skill base.

This particular story has great relevance to other sectors in India and for other developing countries. Entrepreneurs and their supporters need not wait for government policy or institutional reform—they would have to wait a very long time for this. Instead, they should reject any notion that 'development is impossible' because of government bureaucracy and difficulties of doing business. Firms and business associations should be inspired by the Indian case to take the development lead, identify the business opportunities 'out there', and use their creativity to circumvent any barriers to growth. Indian entrepreneurs could do it. So can others.

Notes

1. Other papers focus on the globalization of the Indian software sector (e.g. Bannerjee and Duflo 2000; Desai 2003; Arora and Gambardella 2004; Athreye 2005; Basu 2005). The focus of this chapter is on how entrepreneurial Indian firms managed to enter the industry and grow despite the disadvantages of their economic environment.
2. For an excellent summary of research on this subject, see <www.westaction.org/definitions/def_entrepreneurship_1.html>.
3. See Amsden (1989); Amsden and Hikino (1994).
4. The assumption of an open economy is important. Firms could, of course, avoid the harmful effects of incumbent competition by lobbying for protection but this is a short term gain. In the longer-term, the global market for technology based goods is larger.
5. Estimates from the National Association of Software Services companies (NASSCOM).
6. Big business houses already had previous contacts and relationships with foreign vendors which they exploited.
7. N. R. Narayanamurthy, Chief Mentor of Infosys, has been an important member of many of the corporate governance committees set up for the reform of capital markets in India.

References

Amsden, A. H. (1989) *Asia's next Giant: South Korea and Late Industrialization*. New York: Oxford University Press.

Amsden, A. H., and T. Hikino (1994) 'Project Execution Capability, Organizational Know-How and Conglomerate Corporate Growth in Late Industrialization', *Industrial and Corporate Change*, 3: 111–47.

Arora, A., and A. Gambardella (2004) 'The Globalisation of the Software Industry: Perspectives and Opportunities for Developed and Developing Countries', Working Paper 10538. Cambridge, MA: NBER.

Athreye, S. (2005) 'The Indian Software Industry and its Evolving Service Capability', *Industrial and Corporate Change*, 14 (3): 393–418.

Bannerjee, A., and E. Duflo (2000) 'Reputation Effects and the Limits of Contracting: A Study of the Indian Software Industry', *Quarterly Journal of Economics*, 115: 989–1017.

Bardhan, P. (2006) 'Resistance to Economic Reforms in India', Yale Global Online. New Haven, CT: Yale University.

Basu, I. (2005) 'Global VC Funds Sharpen Focus on India', *The Washington Times*, 2 February.

Baumol, W. J. (1990) 'Entrepreneurship: Productive, Unproductive and Destructive', *Journal of Political Economy*, 98: 893–921.

——— (2005). 'Entrepreneurship and Invention: Toward Restoration into Microeconomic Value Theory', mimeo. New York: New York University.

Beck, T., A. Dermirgüç-Kunt, and R. Levine (2003) 'Small and Medium Enterprises, Growth, and Poverty: Cross-Country Evidence', Policy Research Working Paper 3178. Washington, DC: World Bank.

Cantillon, R. (*c.* 1730) *Essay on the Nature of Trade in General* (translated from French). London: Fletcher Gyles.

De Soto, H. D. (2000) *The Mystery of Capital: Why Capitalism Triumphs in the West and Fails Everywhere Else*. New York: Basic Books.

Desai, A. (2003) 'The Dynamics of the Indian Information Technology Industry', Working Paper 20. London: Centre for New and Emerging Markets, London Business School.

Djankov, S., and C. McLiesh (2005) *Doing Business in 2005: Removing Obstacles to Growth*. Oxford: Oxford University Press, the World Bank and the International Finance Corporation.

Dossani, R., and M. Kenney (2002) 'Creating an Environment for Venture Capital in India', *World Development*, 30 (2): 227–53.

Hausman, R., and D. Rodrik (2003) 'Economic Development as Self-Discovery', mimeo. Cambridge, MA: Harvard University. Available at: <http://ksghome.harvard.edu/drodrik/selfdisc.pdf>.

Hobday, M. (1995) *Innovation in East Asia: the Challenge to Japan*. Cheltenham: Edward Elgar.

Hobday, M., and F. Perini (2005) 'Latecomer Entrepreneurship: a Policy Perspective'. Paper prepared for the Joseph Stiglitz 'Task Force on Industrial Policies and Development' Workshop, held in Rio de Janeiro, Brazil, 17–19 March 2005.

Johnson, S., J. McMillan, and C. Woodruff (1999) 'Contract Enforcement in Transition', Discussion Paper 2081. London: Centre for Economic Policy Research.

——— ——— ——— (2002) 'Courts and Relational Contracts', *Journal of Law, Economics and Organization*, 18 (1): 221–77.

Kapur, D. (2002) 'The Causes and Consequences of India's IT Boom', *India Review*, 1 (2): 91–110.

Khanna, T., and K. Palepu (2000) 'Is Group Affiliation Profitable in Emerging Markets? An Analysis of Diversified Indian Business Groups', *Journal of Finance*, 55 (2): 867–91.

Kirzner, I. M. (1979) *Perception, Opportunity, and Profit*. Chicago, IL: University of Chicago Press.

——— (2000) 'The Limits of the Market: the Real and the Imagined'. In I. M. Kirzner (ed.), *The Driving Force of the Market—Essays in Austrian Economics*. Milton Park: Routhledge.

Knight, F. H. (1921) *Risk, Uncertainty and Profit*. Boston, MA: Houghton Mifflin.

Krueger, A. O. (1974) 'The Political Economy of Rent-Seeking Society'. *The American Economic Review*, 64 (3): 291–303.

Leff, N. (1974) 'Economic Development and Entrepreneurship: the Problem Revisited', *Journal of Economic Literature*, 14: 46–64.

Nair, K. (2004) 'Coming, Silicon Valley VCs', *Business Standard,* 15 December.

North, D. (1990) *Institutions, Institutional Change and Economic Performance*. Cambridge: Cambridge University Press.

——— (2005) *Understanding the Process of Economic Change*. Princeton, NJ: Princeton University Press.

Rothwell, R., and W. Zegveld (1982) *Innovation and the Small and Medium Sized Firm*. Boston, MA: Kluwer Nijhoff Publishing.

Say, J. B. (1816) *Catechism of Political Economy*. London.

Schumpeter, J. A. (1934) *The Theory of Economic Development*. Cambridge, MA: Harvard University Press.

Shadlen, K. C. (2004) *Democratization Without Representation: the Politics of Small Industry in Mexico*. University Park, PA: The Pennsylvania State University Press.

Teece, D. J. (1986) 'Profiting from Technological Innovation: Implications for Integration, Collaboration, Licensing and Public Policy', *Research Policy*, 15: 285–305.

Von der Fehr, N.-H. M. (1995) *The African Entrepreneur: Evidence of Entrepreneurial Activity and Firm Formation in Zambia, Working Paper. Washington DC: World Bank.*

World Bank, International Finance Corporation (2006) *Doing Business in 2006: Creating Jobs*. Washington, DC: World Bank and the International Finance Corporation.

Index

Note: page numbers in *italics* refer to Figures and Tables.